Chemistry and Technology of
Printing and Imaging Systems

Chemistry and Technology
of
Printing and Imaging Systems

Edited by

P. GREGORY
Zeneca Specialties
Manchester

SPRINGER-SCIENCE+BUSINESS MEDIA, B.V.

First edition 1996

© 1996 Springer Science+Business Media Dordrecht
Originally published by Chapman & Hall in 1996
Softcover reprint of the hardcover 1st edition 1996

Typeset in 10/12pt Times by Cambrian Typesetters, Frimley, Surrey

ISBN 978-94-010-4265-9   ISBN 978-94-011-0601-6 (eBook)
DOI 10.1007/978-94-011-0601-6

A catalogue record for this book is available from the British Library
Library of Congress Catalog Card Number: 95–76794

∞ Printed on permanent acid-free text paper, manufactured in accordance with ANSI/NISO Z39.48-1992 (Permanence of Paper).

Preface

Printing and imaging has a major impact on everyone. From the obvious examples of newspapers, magazines and comics through to photographs, currency and credit cards, and even the less obvious example of compact discs, everyone is familiar with the end products of printing and imaging.

Until recently, the major printing and imaging technologies have been impact printing and silver halide photography. Important impact printing technologies are offset lithography, gravure, flexography and screen printing. All these technologies, including silver halide photography, are mature and have changed little over the past few decades. In contrast, the phenomenal growth of silicon chip technology over the past 15 years or so has spawned a new era of printing and imaging systems, the so-called non-impact (or electronic) printers.

Not all the non-impact printing technologies are of equal commercial importance. Some, like diazotype and conventional photolithography, are mature and are declining in importance. Other technologies, though relatively new, have not achieved notable commercial success. Electrography and magnetography fall into this category. The remaining technologies such as optical data storage (the technology used in compact discs), thermography (the technology used in electronic photography), ink-jet printing and electrophotography are the non-impact printing technologies that are both modern and which have achieved remarkable commercial success, especially ink-jet printing and electrophotography. Indeed, electrophotography, the technology which embraces photocopiers and laser printers, together with the more recent ink-jet printers, are now the standard office printers, having displaced the once dominant impact printers such as typewriters, daisywheel and dot-matrix. This revolution has seen printing move from the workshop and darkroom into the office and home.

The advances in non-impact printing technologies are now beginning to be incorporated into some of the mature technologies to improve their performance. Examples include the use of photoconductor technology used in electrophotography, and laser imaging used in optical data storage, to produce lithographic plates, both technologies having advantages over the established photolithography. Indeed, as time progresses there will be more of this interaction of modern technologies with mature technologies.

Because of the rapid changes in the printing and imaging technologies over the last decade, the time is ripe for a book. I am honoured to have been asked to edit such a book and wish to thank all the contributors for their excellent contributions.

Finally I would like to thank my wife, Vera, and two sons, Andrew and Michael, for their perseverance and patience during this project, and Dr Alan Calder, General Manager of Zeneca Specialties, for use of the library and secretarial facilities, especially Amanda Collier for typing several chapters.

Peter Gregory

Contributors

P. Bergthaller   Agfa Gevaert AG, Postfach 10 01 60, D5 1301 Leverkusen, Germany

R. Bradbury   Zeneca Specialties, PO Box 42, Hexagon House, Blackley, Manchester M9 8ZS, UK

R.S. Gairns   Zeneca Specialties, PO Box 42, Hexagon House, Blackley, Manchester M9 8ZS, UK

P. Gregory   Zeneca Specialties, PO Box 42, Hexagon House, Blackley, Manchester M9 8ZS, UK

P.A. Hunt   ICI Imagedata, Brantham, Manningtree, Essex CO11 1NL, UK

T.S. Jewitt   Department of Communication Media, Manchester Metropolitan University, Cavendish Street, Manchester M15 3BR, UK

R.W. Kenyon   Zeneca Specialties, PO Box 42, Hexagon House, Blackley, Manchester M9 8ZS, UK

Contents

5 Ink jet printing   113
R.W. KENYON

6 Thermal printing   139
R. BRADBURY

7 Optical data-storage systems   168
P.A. HUNT

8 Electrostatic, ionographic, magnetographic and embryonic printing technologies   195
P. GREGORY

1 Setting the scene

P. GREGORY

1.1 From maturity to mayhem

Never since its invention by the Chinese in AD 1041 and its introduction to
the Western world in AD 1450 by Gutenburg has printing experienced such
a turbulent time as it has over the last decade. For many centuries the
fundamental printing processes remained essentially unchanged: advances
came about by the application of improved technology to the processes.
This mature calm has been shattered in the last decade or so! Novel
printing and imaging technologies have burst on to the scene; these are
fundamentally different to the old, mature technologies. Some are so good
in terms of performance and cost that they are displacing, indeed
obliterating, some established printing and imaging technologies. Electrophotography, the technology which embraces photocopiers and laser
printers, and ink jet printing, fall into this category. Photocopying is the
undisputed technology for the reproduction of documents, especially
black-and-white documents, in the office. Laser printers, together with the
more recent ink jet printers, are now the standard printers for the office
and home, having displaced the once dominant impact printers such as
typewriters, daisywheel and dot matrix. It was estimated that in 1994 there
were a total of *c.* 55 million photocopiers/laser printers (*c.* 40 million
photocopiers and 15 million laser printers) and *c.* 16 million ink jet printers
worldwide. Ten years earlier, there were virtually no ink jet printers or
laser printers! Indeed, in a comprehensive review of printing processes in
1982 (Bruno, 1982), out of the 53-page article only two paragraphs were
devoted to laser printing and ink jet printing!

There are two main reasons for this upheaval and rapid change in the
printing and imaging technologies. One is the impact of the silicon chip.
The second is the environment.

1.2 The ubiquitous silicon chip

The first and most important reason for the present mayhem is the
invention of the silicon chip. This tiny device is at the heart of every
semiconductor which, in turn, is at the centre of every electronic gadget

from computers to lasers. Advances have been so rapid in semiconductor technology and miniaturisation that even today computers soon become obsolete. Semiconductors are now so prevalent in our society that future historians may well refer to the current era as the 'Electronic Age'. This semiconductor technology is vital to the two most successful new printing and imaging technologies, electrophotography (photocopying and laser printing) and ink jet printing. It is also essential to optical data storage, the technology behind compact discs.

In addition to producing completely new technologies, silicon chip technology has been used to modernise existing technologies. A good example is the production of lithographic printing plates using either laser printing or laser thermal imaging (the technology used in compact discs). Indeed, how fundamental new discoveries such as silicon chip technology have affected the printing and imaging technologies is a major theme of this book.

1.3   The environment

The second reason for the mayhem concerns protection of the environment. In recent years there has been a strong realisation that we cannot continue to pollute our planet with industrial and domestic waste as we have in the past. Consequently, several international programmes are now in place to limit the discharge of pollutants, especially chemicals, into the environment. As far as the printing and imaging industry is concerned the main areas are the colorant, the solvent and the substrate.

The main problems with colorants have been safety (from a toxicological viewpoint) and disposal. The few colorants based on the carcinogens benzidine and 2-naphthylamine are no longer manufactured. Currently, the aim is to reduce the discharge of heavy metals into the environment. Colorants, especially pigments, containing toxic heavy metals such as lead, cadmium and mercury, have been banned for many years. Now chromium, cobalt and even copper-containing colorants are coming under the environmental spotlight. Although current colorants containing these metals may continue to be used if they are safe, it is likely that new colorant manufacture will try to avoid their use.

The emission of organic solvents into the environment has come under close scrutiny following the effect of chlorofluorocarbons (CFCs) depleting the ozone layer. This has led to international agreements to drastically reduce the emission of organic solvents into the atmosphere. In the printing industry this is causing a shift away from solvent-based processes. In some cases this is to a water-based process or in other cases to a different technology.

Paper is the dominant substrate for printing and imaging. Despite hopes to the contrary, the 'electronic office' has intensified the usage of paper. The demand for paper became so large that additional ways of obtaining paper, other than continually chopping down forests, were sought. This heralded the recycling of waste-paper era so that today recycled paper is becoming more prevalent. Indeed, recycled paper is used as a selling point for products containing it! Challenges for the printing and imaging industries are to print on recycled paper (which is often of inferior quality to virgin paper) and to facilitate the recycling of printed paper products, for instance, to assist the recycling process by making the troublesome de-inking process easier.

1.4 Themes and focii

In conclusion, the focus of this book will be on the modern printing and imaging technologies, particularly the successful non-impact printing technologies, and on the influence and interaction that these technologies are having, both on improving the performance of, and also of competing with, the established technologies such as offset lithography. The emphasis is on the small-scale production of hardcopy output and we will critically compare and contrast the various technologies against such important parameters of cost, speed and performance.

The book consists of nine chapters. Following this brief introductory chapter are chapters on the traditional impact printing technologies and silver halide photography. The remaining chapters represent the main thrust of the book, namely the rapid development and current status of non-impact printing. A chapter each is devoted to the three main technologies of electrophotography, ink jet printing and thermal printing plus a smaller chapter on optical data storage. A further small chapter covers other less important technologies such as magnetography, electrography and ionography. The book ends with a small chapter on future trends.

References

Bruno, M.H. (1982) Printing Processes, in *Encyclopedia of Chemical Technology*, 3rd edn (eds R.E. Kirk and D.F. Othmer), John Wiley, New York, Vol. 19, pp. 110–163.

2   Traditional impact printing
T.S. JEWITT

2.1   Introduction

Almost everyone takes printing for granted, but few people can explain how an image gets on to a substrate, whether it is a magazine, a newspaper, a drinks can, wallpaper or a credit card. Printing has been part of our everyday lives since we were first able to recognise objects, and has influenced our lives in a number of significant ways, e.g. books used in school and university, fashion magazine illustrations, reproductions of works of art and product advertising, whether it be on a hoarding or a package on a supermarket shelf. Printing is the sixth largest industry in the UK with a multibillion pound turnover. When this is allied to its associated activities such as paper-making, ink-making, publishing and packaging then it becomes an enormous multinational industry. This UK industry employs over three million people in some 18 000 companies. It can be argued that the spread of literacy created the need for printing or that printing was responsible for the increase in literacy which helped to spawn the Renaissance.

2.2   A brief history of printing

In the 15th century, even with an army of monks working all the daylight hours, it took many days to make a fair copy of a book. Pictures were laboriously painted by hand and were extremely costly to produce.

Around 1450, Johann Gutenberg of Mainz in Germany developed a technique for casting individual letters or types from moulds. These types could be assembled to form pages, which could be inked and, when a sheet of paper was pressed on them, produce a full page of 'writing' in a minute or so. This invention of moveable type revolutionised the hitherto crude process of printing. The monks in the scriptorium were rapidly to go out of business because of this, pre-dating the Industrial Revolution by 300 years. The invention was developed steadily over the whole of Europe, becoming an industry on which civilisation depended for the dissemination of knowledge and information.

It was not until 1800 that the first all iron press was invented. However, the process of printing remained essentially as it was when Gutenberg used it. The Industrial Revolution played an enormous part in the development of printing. The cylinder press was developed, whereby the sheet of paper was attached to a cylinder which rolled over the type to be printed. By the application of steam and later electricity, printing speeds were increased from 50 or so impressions per hour to many hundreds per hour. By the middle of the 19th century the rotary press had been developed, with the type in the form of a cast 'stereotype' fixed on to a cylinder. The paper in reel form was pressed on to the stereotype by an impression cylinder. This produced press speeds of 25 000 impressions per hour. Automatic typesetting machines had also been developed to produce either separate types or lines of letters cast on one body. This method of printing from raised characters 'letterpress' dominated the first 500 years of printing.

By the beginning of the 19th century, printing was developing with other processes. Intaglio printing had been used to reproduce illustrations for many years. Copperplate engraving had been in use since 1446 with some of the finest work being produced by Albrecht Dürer (1471–1528). The artist incised the image into a copper plate, the image was filled with ink and damp paper pressed on to the surface of the plate by means of a heavy roller. The ink was pulled out of the incisions on to the paper to produce the print.

'Etching' was also an intaglio process. A soft 'resist' was coated on to the metal plate and the image created by the artist scratching the design through this resist. The plate was then etched in a mordant to a depth sufficient to hold the ink. Printing was undertaken in a similar way to a copper plate engraving. The processes of letterpress and intaglio printing developed side by side. Letterpress brought literature to a mass audience whilst intaglio provided art, music, maps and illustrations to this same mass audience.

Rotary intaglio had been used for wallpaper and calico printing since 1785, but it was Karl Klietsh, a Czechoslovak working in Vienna in 1886, who developed the modern gravure process. Moving to England he joined Storey Brothers of Lancaster who were calico and oil cloth printers. Working together with Samuel Fawcett, an engraver at Storey Brothers, they developed a method of combining photography and engraving which is still in use today.

In the late 18th century, Johann Aloys Senefelder was experimenting in Munich to produce a cheaper method of printing, either by relief or intaglio. Legend has it that his famous discovery of lithographic printing (*lithos*, a stone; *graphis*, to write) was accidental, when a shortage of paper forced him to write his laundry list to his landlady on a piece of polished limestone with a wax crayon. He found he could not remove the image from the stone, but could print from it. He patented the process in Britain in 1800. It is based on the fact that grease and water are mutually repellent; the design was drawn on flat polished limestone in a greasy ink. The stone was then dampened with water and a printing ink applied to the surface.

This ink was repelled from the non-image areas which were wet, but adhered to the greasy image. A sheet of paper was then laid over the stone and line pressure applied to transfer the image to the paper. Exceptionally fine colour work was produced by this method, Toulouse Lautrec being one of the finest lithographic artists. Senefelder, being a poor artist himself, invented a drawing machine; he also developed a method of drawing on to paper and transferring the image to stone. This gave rise to the first multi-image duplicating process. He also discovered that prepared zinc plates could be used instead of stone. Though the trade were slow in taking up the idea, zinc plates dominated lithography until the 1960s when they were generally replaced by aluminium as the image carrier. Offset lithography was patented in 1875 for printing on tinplate and has steadily developed to become the dominant printing process in use today.

Around the end of the 19th century, a simple method of relief printing had been invented in Liverpool by Bibby and Baron, called aniline printing. It was a high-speed rotary web-fed process. The image to be printed was cut in rubber, which was then attached to a cylinder for printing. The inks used were liquid and based on aniline dyes. The quality of the printed image was very poor when compared to those of other methods available, and despite many experiments, was confined to what it had been designed for, namely the printing of paper bags. The simplicity of the process had attractions and was developed, mainly in the field of package printing, for cartons and Cellophane. By the 1950s, the process was considerably improved and had changed its name to 'flexography' as inks other than the original aniline alcohol inks were now in use. This process is now the dominant relief printing process and is still developing rapidly.

Probably the single most significant influence on the development of printing in its first 500 years was the invention of photography in France, in the early part of the 19th century. Printed illustrations by relief printing were made from wood engravings. These had largely replaced 'wood cuts' which were comparatively crude in appearance. The engravings and etchings produced by intaglio printing and the illustrations produced by lithography were very much in the hands of creative artists. Many attempts to marry this new imaging technology, 'photography' with printing were tried with very few successes, until in 1886 F.E. Ives invented the cross-line screen. This device, placed in the optical path of a camera, split up the image into dots of varying size corresponding to the tones of the original. From this 'half-tone' negative, printing surfaces could be made for all the printing processes, although it was not until the 1920s that a successful application of the half-tone process to gravure was developed.

The development of the microchip and its application to imaging systems has brought about a greater rate of change in printing technology in the last 25 years than had ever been seen before. Printing has developed from an art form, via a craft, to one of the world's leading 'high tech' industries. In any survey of the history of printing, consideration should be given to the methods of 'printing' used in business and commerce. The 15th century had seen the demise of handwritten books; however, legal documents and correspondence continued to be written by hand, with the possibility of taking one copy by means of dampened paper in a heavy press.

Mechanical writing machines were proposed as early as the 1700s, but it was not until 1868 that the first workable machine was patented by Glidden, Scholes and Soule in Milwaukee, USA. This machine was refined by Christopher Scholes and manufactured by E. Remmington and Sons. It appeared on the market in 1874. The typewriter was slow to be adopted by business and commerce, but by 1900 there were at least forty manufacturers of typewriters worldwide. The type was on bars, with upper- and lower-case characters, the letter struck an inked ribbon and was impressed on the paper when a key was struck. There was only one typeface available on any one machine and each letter occupied the same amount of space. The 'QWERTY' keyboard was designed to prevent the keybars sticking to each other when adjacent characters were struck on the keyboard at speed. Electric typewriters were invented in the 1920s and in 1933 Varityper incorporated a system of mounting the type on a semicircular metal plate which was interchangeable, but it was not until 1946 that differential letter spacing was introduced by International Business Machines (IBM) who in 1961 marketed the first 'golf-ball' typehead.

A good typist would produce between 80 and 100 words per minute on a manual typewriter and between 110 and 120 words per minute on an electronic typewriter.

The typewriter has developed via the application of electronics to have memory systems and this leads to word processing where the information is keyed in and stored on magnetic media. The print-out speeds are in the order of 8–10 times faster than that of a typewriter.

2.3 Printing technologies

2.3.1 Relief printing

In relief printing, the image area or printing surface is raised above the non-printing areas (Figure 2.1). Ink is applied to the printing surface by rollers and is transferred to the substrate by means of pressure. The ink film thickness is controlled in order to produce a sharp edge to the image.

2.3.2 Letterpress

This process is the oldest of the printing technologies. The images are printed from metal type, cast in an alloy of lead, tin and antimony and held

Raised image areas only receive ink and print

Non-image areas not in relief do not receive ink and therefore do not print

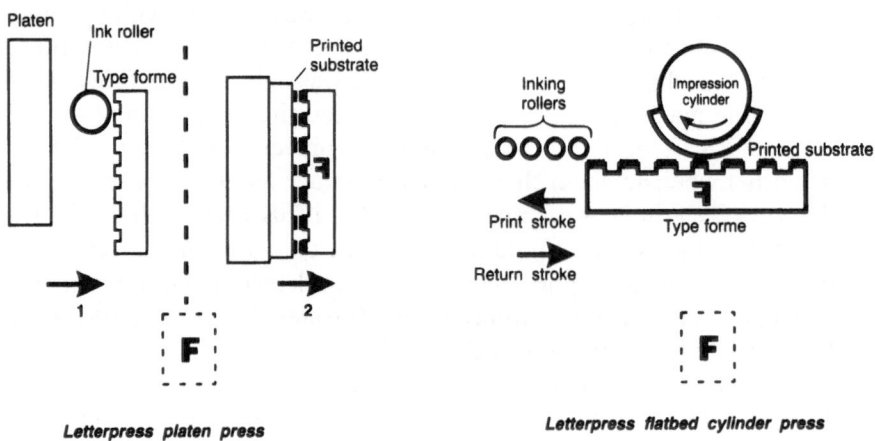

Letterpress platen press

Letterpress flatbed cylinder press

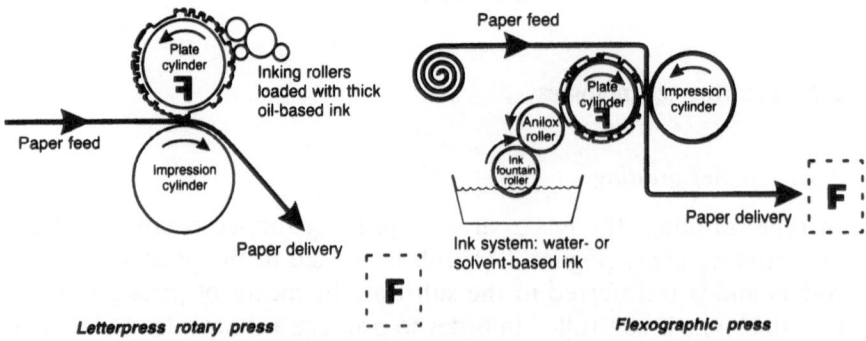

Letterpress rotary press

Flexographic press

Figure 2.1 Relief printing. Reproduced from *Introduction to Printing Technology* (4th edn) with permission from the British Printing Industries Federation.

in page format along with any illustrations by means of a metal frame. Illustrations were produced separately on to copper or zinc plates. A photographic process was used to obtain an image on the metal which was then chemically etched into a relief format. The process was known as photo-engraving. The 'forme' of type and illustration was mounted on a flat-bed press in either a vertical or horizontal position for printing. Rotary letterpress originally used metal stereotypes cast in a half round format from papier mâché moulds made from the 'forme' of type and illustrations. Modern rotary letterpress use a hard photopolymer plate as the printing surface, which is produced by a photographic process.

Letterpress printing inks are of a group known as paste inks. They are composed of a varnish, additives and pigment. The varnish was traditionally a hard resin such as a rosin-modified phenolic resin dissolved in an oil, normally a vegetable oil such as linseed oil, which will dry by oxidation/ polymerisation. Latterly, the drying oil has been replaced by long-chain alkyd resins to which phenolic, maleic or hydrocarbon resins are added. The additives used in printing ink formulations are to control such properties as on-press stability, drying rates, rub resistance, etc. The choice of pigment will depend on the colour required. Black pigments are all carbon blacks, whilst the coloured pigments used are organic and very wide ranging. Inorganic whites are used both as a pigment or as an extender, with titanium dioxide being the most commonly used.

The inks, when mixed and milled, have a high viscosity and in use on the press require a complex train of rollers to break the ink film down to the thickness required for printing.

2.3.3 Flexography

Flexography is relatively modern and is a highly successful web-fed printing process. The relief printing surface is usually of a soft rubber or photopolymer, mounted on to a plate cylinder. Liquid inks are used and these dry mainly by solvent evaporation. Traditionally, these were organic solvent-based, but increasingly, water-based inks and ultraviolet-cured methods of ink drying are being used.

The flexographic process is used to print on a wide variety of substrates from paper and board to impervious substrates such as plastic films, e.g. polyolefin, polypropylene, polyvinyl chloride etc. and metals and metallised foils. The inks used will vary in formulation to suit the substrate and end-user requirements for the printed product. Additionally, the materials for the printing surface will influence the type of solvent used in the ink formulation. The printing surface can be produced by a number of methods using different materials. Traditionally, a stereotype plate was produced from metal type and illustrations, in much the same manner as that used for rotary letterpress. The mould of the matter to be printed is made by pressing a thermosetting resin material on to the printing forme in a heated hydraulic press. This produces a hard, rigid mould of the type and illustrations. The printing stereo is produced by pressing a sheet of uncured rubber into this mould again in a heated hydraulic press. This produces a copy of the original metal type, etc. in flexible rubber, which, after grinding the back to ensure even thickness, can be mounted on to the press printing cylinder with a double-sided adhesive.

An alternative to the rubber stereo is to engrave the printing surface directly into rubber by means of a laser. Separate sheets or complete rubber rollers can be engraved, for text and line or half-tone illustrations. Half-tone up to 60 lines/cm (150 lines/inch) can be produced by this method.

The most common method of printing surface preparation for flexography in current use is the photopolymer plate. Plates are available in a variety of thicknesses and hardness on a wide range of backing material.

The photopolymer material, which may be solid or liquid, is exposed to high-intensity ultraviolet light through a photographic negative. Where the light penetrates the clear image areas of the negative, the material polymerises and becomes insoluble in the 'developer'. The non-image areas, which are not polymerised, are removed in a plate-processing machine by the use of a solvent, usually water-based. In the case of liquid photopolymer plates, an air knife is used to remove the unpolymerised material which is then recycled into the system for re-coating.

Flexographic inks consist of a colorant, binder and a solvent. The colorants traditionally used in flexographic inks are the salts of cationic dye bases. Other light-fast dyes are used which are mainly metal complexes. However, dye-based inks are not widely used except for paper bags, waxed food wrappers and decorative papers such as wrappings, napkins, kitchen towels, etc.

Pigments are more commonly used and these have similar specifications to those used in other printing inks. The dispersion properties in the various solvents used and the wettability of the pigment will largely govern the choice of pigments for use in flexographic inks. The binder will be a resin, which in addition to binding the colorant to the printed substrate also acts as a carrier for the dye or pigment in the liquid state of the inks. The choice of resin will be governed by its solubility in types of solvent chosen to have no effect on the material used for the printing surface. Amongst the resins used are nitrocellulose, ethyl cellulose, polyamide resins, acrylates and methacrylates.

The solvent is the carrier which makes the ink a liquid, but it must be easily removed from the substrate by evaporation (and possibly some penetration) as soon after printing as possible. The most common solvents used are ethyl, isopropyl and n-propyl alcohols and water. Additionally, ethyl, isopropyl and n-propyl acetates, aliphatic and aromatic hydrocarbons or ketones may be used for specific types of ink.

2.3.4 Intaglio printing

Apart from some highly specialised techniques used in bank-note printing, commercial intaglio printing (Figure 2.2) is confined to the gravure process. In the gravure process, the image of type and illustrations is recessed into the surface of the image carrier. Gravure is predominantly a reel or web-fed process and the image carrier is usually a copper cylinder, the surface of which is chrome plated to give better wear characteristics. The image is in the form of discrete incised cells, which in the conventional photogravure process are of constant area but variable depth. The deeper the cell, the more ink is transferred to the paper. Tone differences are achieved by controlling cell depths at the etching stage. Methods of varying both cell area and cell depth have been in use since the 1920s by the use of special half-tone screens to produce what is known as invert half-tone gravure.

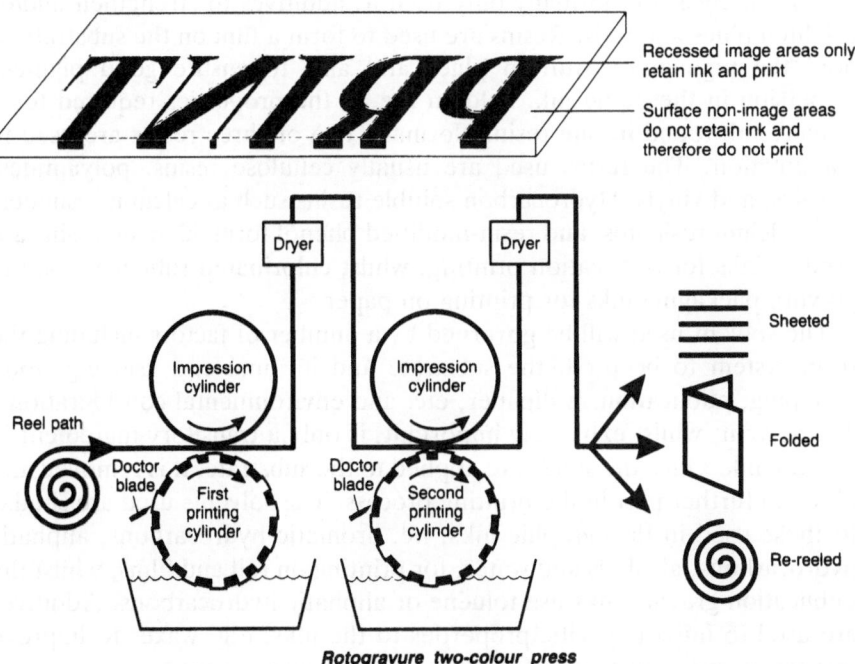

Rotogravure two-colour press

Figure 2.2 Intaglio printing. Reproduced from *Introduction to Printing Technology* (4th edn) with permission from the British Printing Industries Federation.

In the modern gravure process, the cylinders are predominantly produced by electromechanical engraving. The originals to be reproduced are scanned on a dedicated drum scanner. The processed information drives multiple diamond tooled engraving heads to cut the cells into the copper cylinder. Alternatively, digital information direct from CEPS (Colour Electronic Pre-press Systems) is used to control the engraving machine. Printing by this process consists of a number of printing units in line. Each unit is made up of an ink duct, in which the prepared cylinder rotates in a liquid ink. A finely ground steel blade ('the doctor blade') in contact with the cylinder reciprocates sideways in order to remove the excess ink from the surface of the cylinder. The web of substrate passes into a nip between the inked printing cylinder and a rubber surfaced impression cylinder. The ink is transferred from the printing cylinder by pressure alone or by pressure plus electrostatic forces. The printed web then passes into a drying system (where the solvents are removed by evaporation and recovered for re-use), and on to the next printing unit.

From the above brief description of the gravure printing process it will be apparent that the requirements of gravure printing inks are mobility, low viscosity and very rapid drying. Gravure inks contain colorants, resins, additives and solvents. The colorants are usually organic pigments; however, light-fast dyes are sometimes used when printing on aluminium foil. Basic dyes are normally only used as additives to strengthen and/or brighten blues and reds. Resins are used to form a film on the substrate to hold the pigment/dyestuff in place and also to ensure good pigment dispersion in the liquid ink. Seldom are all the properties required for a gravure ink found in one resin. Normally two or three resins are used in combination. The resins used are usually cellulose resins, polyamides, acrylics, and vinyls. Hydrocarbon soluble resins such as calcium resinates, zinc-calcium resinates, and resin-modified phenol formaldehyde resins are used in inks for publication printing, whilst chlorinated rubber is used in gravure packaging inks for printing on paper.

The solvent used will be governed by a number of factors including the resin system to be used, the substrate and its final end use, e.g. food wrapping, publication, wallpaper, etc. and environmental considerations. The solvent, whilst extremely important, is only a transitory ingredient in gravure ink. Once the solids are applied to the substrate, it is removed and plays no further part in the printing process. The solvents used are similar to those used in flexographic inks, i.e. aromatic hydrocarbons, aliphatic hydrocarbons, alcohols and water, for printing on foil and films, whilst the publication gravure inks use toluene or aliphatic hydrocarbons. Additives are used to impart specific properties to the inks, e.g. waxes to improve resistance and optical brighteners to improve appearance.

2.3.5 Planographic printing

There are three excellent methods of printing based on the planographic process, that is where the image and non-image areas are virtually in the same plane. The first, collotype, is a photographic process where the printing surface is reticulated gelatine. This is a short run, slow printing process. It is very difficult to control due to the delicate nature of wet gelatine. It is now only very rarely used for fine art reproductions and will not be considered further.

The second method, screen printing, is a form of stencil printing (Figure 2.3) where a thick viscous ink is forced through the image area of the stencil on to the substrate. Originally called 'silk screen printing' because silk was used as the mesh stencil carrier, in modern screen-process printing the mesh for the stencil is usually nylon, polyester or stainless steel. This material is stretched taughtly over a frame before imaging. The image may be hand-cut into an adhesive foil which is then positioned on the mesh material, or produced by a computer-controlled stylus. Photography is also employed, where the image is generated by a photographic process and then transferred to the stencil material or, alternatively, the mesh material is coated with a photopolymer, exposed to a photographic positive and the unpolymerised coating removed, leaving a stencil for printing. Printing, which may be by hand, semi- or fully automatic is achieved by placing the substrate under the stencil frame holding the image. The frame is 'loaded' with ink and by the use of a squeegee, the ink is forced through the clear parts of the stencil on to the substrate. The ink may be allowed to dry naturally by racking the printed sheets or an accelerator such as ultraviolet radiation may be used.

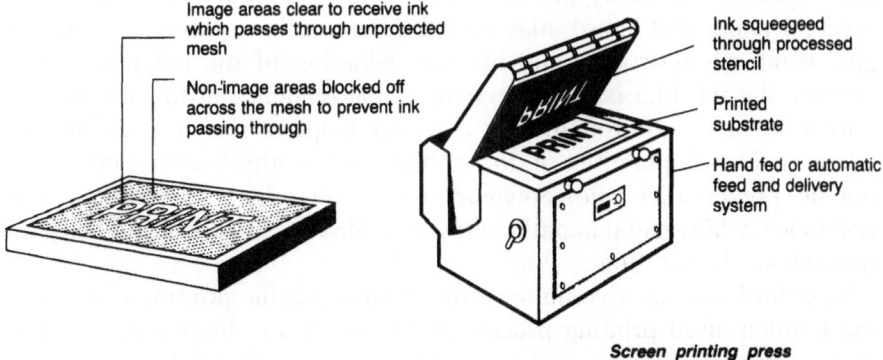

Image areas clear to receive ink which passes through unprotected mesh

Non-image areas blocked off across the mesh to prevent ink passing through

Ink squeegeed through processed stencil

Printed substrate

Hand fed or automatic feed and delivery system

Screen printing press

Figure 2.3 Stencil printing. Reproduced from *Introduction to Printing Technology* (4th edn) with permission from the British Printing Industries Federation.

The thickness of the ink film is a characteristic of this process although this is not always apparent in textiles printed by screen printing. Screen-printing inks may be opaque or transparent, depending on end-user requirements and whether the half-tone process is being used. The composition of screen inks is common to most other printing inks, i.e. pigments, resins, solvents and additives; however, many of the requirements are peculiar to screen inks.

End-user requirements are especially important when choosing pigments. Much of screen-printed material is used for display work, therefore light-fastness is extremely important, as is weather-resistance. Most organic reds do not meet these requirements; quinacridone red pigments are the exception and therefore tend to dominate in screen ink formulations.

The consistency of the ink is extremely important and can be influenced by the choice of extender. Calcium carbonate or china clay are frequently used.

Pigments must be compatible with the solvents to be used, for example PTMA (phosphotungstomolybdate) pigments would be unsuitable for use with glycol ethers, ketones and esters. A wide range of solvents may be used in screen inks. Because the screen ink is open to the atmosphere during use in the squeegeeing process, the solvent evaporation rate does to some extent limit choice. The solvent selected must keep the ink fluid during printing, but must also evaporate easily and quickly during drying on the substrate. 'Cosolvent inks' are frequently used; usually these are mixtures of propylene glycol ethers and aromatic or aliphatic hydrocarbons.

The solvent should also have the minimum of effect on the polymers of photographic stencils or on the material which constitutes the squeegee. The choice of resin will be to a large extent governed by the end use. However, the resin chosen must have the common properties of film forming with a tack-free surface and because screen inks dry by solvent evaporation, the resins must release the solvent freely. Acrylic resins have good adhesion to many plastics and cellulose-derived resins are widely used for paper and board inks. Additives are necessary in screen inks to give good rub resistance and increase adhesion of the ink film. Also, because the ink film is often very thick, the tendency of this ink film to bubble must be controlled. The extender helps in this respect but the addition of small quantities of, for example, polymethysiloxane will act as a bubble preventative. Polyethylene waxes are added to enhance rub resistance, whilst metal naphthenates are added to ensure thorough drying throughout the ink film.

The third and most common form of planographic printing is also the most widely used printing process in the world, i.e. lithography. Lithographic printing (Figure 2.4) or more properly, 'offset lithography' has come to dominate conventional printing due mainly to its speed, versatility, quality and cost effectiveness as a producer of multiple hard

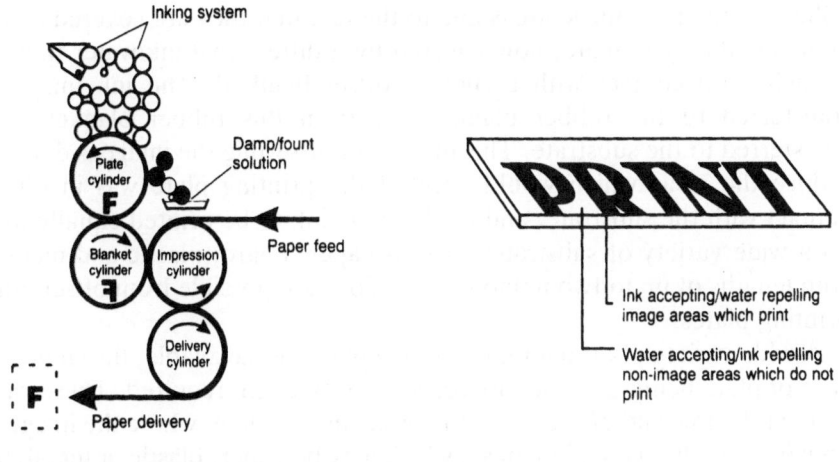

Offset litho single-colour printing press

Figure 2.4 Lithographic printing. Reproduced from *Introduction to Printing Technology* (4th edn) with permission from the British Printing Industries Federation.

copy images in black and white or multicolour. The image and non-image areas are virtually in the same plane on the printing image carrier. The image areas are hydrophobic and the non-image areas are hydrophilic. This is brought about by the selective chemical treatment of the printing plate (which is usually aluminium) to ensure different surface energies between the image and non-image areas.

The surface of the aluminium plate is normally anodised. This produces a hard porous layer on the surface of the plate which increases the hydrophilic properties of the metal. A light-sensitive coating of either a diazo, or more usually a photopolymer, is applied to this anodised layer. After exposure to ultraviolet light through either a photographic negative or positive, the light-sensitve coating undergoes a change of state of solubility and the non-image areas of the coating are removed by means of a solvent (usually water or water-based). The remaining coating is in the form of the image and is oleophilic (hydrophobic). The surface of the plate is treated with a 'finisher' such as a gum arabic solution to ensure the complete desensitisation to grease of the non-image areas and to preserve the surface from any chemical attack prior to printing.

The plate is positioned on the plate cylinder of the press and damped with an aqueous 'fount' solution. This forms a thin aqueous film over the non-image areas, whilst on the hydrophobic image areas, the surface energy is such as to prevent a film forming. The aqueous fount solution which initially covers the image area contracts into tiny droplets; these are displaced by the ink film which is then applied to the plate from the inking rollers. The ink film adheres to the oleophilic image areas, but is unable to adhere to the non-image areas due to the fact that they are covered by an aqueous film. The plate, now covered by a differential ink/water film, is brought into contact with a rubber 'offset blanket'. The ink image is transferred to the rubber blanket and from this rubber blanket it is transferred to the substrate. This process of offsetting the image is done to reduce the wear which would occur if the printing plate was in direct contact with the substrate, and to allow the ink to be printed equally well on a wide variety of substrates such as paper, board, plastic and metals. Run lengths of up to two million impressions are possible from offset litho printing plates.

A wide variety of lithographic printing plates are available, the choice of type of plate being governed by costs and run length required. These vary from at the low end of the scale, the direct image plate, where the image is applied directly to the plate base, which may be paper, plastic or metal, by for example, a greasy typewriter ribbon. These plates are cheap and quick to produce and will print 400–500 impressions. At the other end of the scale are multimetal plates. In these plates, a hydrophilic metal such as chromium, stainless steel, or aluminium is used for the non-image areas and an oleophilic metal such as copper or one of its alloys is used to form the image areas. These plates are very expensive to produce but are capable of run lengths in excess of two million impressions of very high quality.

2.4   The offset printing press

The configuration of the cylinders on the press as shown in Figure 2.4 may differ somewhat in the various types of press in use, i.e. single colour, multicolour, perfector, sheet- or web-fed. However, the fundamental principles of plate cylinder, blanket cylinder, impression cylinder and delivery system are common to all presses.

2.4.1   Image quality

The quality of the printed image is a combination of the control of a number of factors, namely the image on the plate, the offset blanket, the fount solution, the ink and the substrate.

2.4.2   The image on the plate

As has been previously stated, the image may be created on the plate by simple direct means or more commonly by a photographic process, whereby the light-sensitive coating on the plate is exposed through a photographic negative or positive to ultraviolet light to create an image

(Gregory, 1994). Other techniques are being used and these will be discussed later.

2.4.3 The offset blanket

The blanket is the surface which transfers the ink to the substrate in lithography, making it responsible for much of the quality of the printed image. The blanket is composed of laminated layers of woven cotton or synthetic fibres which in turn are coated with a synthetic rubber or other similar compound, the whole having a smooth surface and overall equal thickness. Offset blankets are manufactured in rolls, usually 50 metres in length, in varying widths in excess of 1 metre. Surface grinding techniques produce a product where the thickness variation is only 0.03 mm along a 50 m length and 0.02 mm across the width. The surface materials are chosen to give properties required on the press such as hardness, effects of ink composition, solvent resistance, etc. Compressible blankets are used for most types of printing today; these give better image edge definition, have lower stretch characteristics and are more resistant to smash and edge cutting. The compressible characteristic is given by incorporating a layer of air or gas bubbles in the structure. This enables the structure to deform without stretching.

There is no universal blanket because of the wide variety of substrates and numerous diverse printing applications in offset lithography.

2.4.4 Fount solution

A controlled amount of fount solution is applied to the surface of the printing plate at each revolution of the press by means of the damping roller train which is in contact with the plate prior to the inking rollers (Figure 2.4).

The major function of the fount solution is to keep the non-image areas wetted with the minimum amount of water, which prevents the ink from being deposited there and thus ensures a clean print on to the substrate. The fount solution also continuously replaces the desensitising layer on the non-image areas. Apart from water, the desensitiser is the main ingredient in fount-solution formulations. Traditionally, gum arabic was used as a desensitiser, but since the introduction of isopropyl alcohol into fount solutions synthetic alternatives to gum arabic have had to be used, due to the incompatibility of gum arabic with alcohols. Fluorides, phosphates, nitrates or phosphate esters may be employed as alternatives to gum arabic in these alcohol-containing solutions. Fount solutions are complex formulations: besides the desensitiser, a pH buffer is required to hold the pH between 5 and 7 and citrates or phosphates are used for this purpose. The use of isopropyl alcohol to reduce the thickness of the water film on the plate is environmentally unacceptable and is rapidly being replaced by anionic or block co-polymer surfactants. Polymeric surfactants or glycols are used as humectants in order to prevent the plate surface from drying too readily when the press is temporarily stopped. Water softeners, biocides, corrosion inhibitors and anti-foaming agents are also included.

2.5 Lithographic inks

The lithographic printing process embraces a wide variety of materials and techniques, each needing specialised ink formulations. Therefore, only a generalised treatment of lithographic inks is possible in this chapter. Lithographic printing inks are 'paste' inks and are composed of varnishes, additives and pigments.

The choice of the formulation of the varnish and additives used will depend on the printing process, subtrate and drying mechanism. Inks used in lithographic printing dry by the well-understood methods of penetration, quick-setting phase separation, oxidation polymerisation, heat setting or radiation curing.

The ink film used on the press is generally between 1 and 3 μm in thickness, therefore a high pigment concentration, with excellent dispersion properties is required in order to achieve satisfactory colour strength on the substrate.

As has been previously stated, in most lithographic printing processes, water is present on the printing plate. Therefore, a degree of emulsification is required to remove this water/fount solution from the image areas. This property must be carefully controlled in order to maintain the basic principles of lithography. Too much emulsification would destroy the balance between ink and water, which would cause the plate to 'go blind' and the non-image areas would print. Therefore, materials are included in the ink formulation which will combine with the fount solution on the inked image and keep the plate clean. Lithographic printing inks can be classified into groups which reflect the printing technique and materials used, coupled to the ink drying mechanism.

2.5.1   Cold-set web-offset inks

These inks are printed on to an absorbent substrate such as newsprint and 'dry' by penetration. In fact, the ink is trapped between the fibres of the substrate and does not dry. This system is widely used in newspaper and book production as it has advantages of very high printing speeds at low costs. The formulation may be a pigment, a varnish, mineral oil or soya oil and additives.

2.5.2 Heat-set web-offset inks

After printing, the web passes into a drying oven which uses hot air to raise the temperature of the ink to around 120–140°C in stages. This serves to evaporate the fount solution in the ink or web without boiling out and disrupting the ink film. At the same time, the heat removes 85–90% of the distillate in the ink. The web then passes over chilled rollers to reduce the web temperature to ambient before it enters the folder and trimmer.

This ink-drying method produces a high-quality gloss image on even quite low-quality papers, and is very widely used for printing magazines, brochures, leaflets etc. where runs of 20 000 or more are required. The ink formulation is complex because of the high demands made on the ink during printing and drying at very high speeds. It will contain a pigment, a vehicle system (consisting of hard resins such as rosin-modified phenolics or hydrocarbon resins, rosin esters or phenolic-modified hydrocarbons; alkyds, as pigment wetting vehicles; distillates, as both solvent and diluents – these are normally high boiling point petroleum distillates; gelling agents, to increase the viscosity and yield value, normally polymeric organic aluminium compounds) and additives, such as waxes, to provide slip in the ink during drying and cooling, and rheology modifiers such as aluminium chelates, clays or silicas. Supersolvents may also be added to make the resin system more soluble, e.g. higher alcohols such as tridecyl alcohol.

2.5.3 Sheet-fed lithographic inks

Good-quality book work, calendars, brochures, general printing, packaging and label printing on paper and board use this method of printing, as does metal decorating. The inks, however, are somewhat different for paper and board than those used for metal decorating.

The drying process for all these inks is a three-stage process. Firstly, the distillate penetrates the surface of the substrate very rapidly. The high-viscosity materials remain on the surface and 'set' due to increased viscosity, at the second stage. This may take from 2 to 40 minutes depending on the ink formulation and the surface of the substrate. The ink is sufficiently dry and hard at this stage to prevent 'set off' or marking the back of subsequent sheets in contact with them during printing. The third stage is the aerial oxidation of the set film. This forms a hard, dry, flexible ink film due to the polymerisation of the drying oil or alkyds. This is a relatively slow reaction taking 6–12 hours before the sheets are usable for further processes and some days before the reaction is complete.

The inks are composed of a pigment and a vehicle – hard resins, drying oil or alkyd plus solvents such as petroleum distillates and additives. These are normally driers, to accelerate the oxygen-induced polymerisation process (these are organic salts of magnesium, cobalt, zirconium etc.), anti-oxidants, to prevent surface drying of the ink in the duct on the press (these are usually quinones but may be substituted phenols) and waxes, to produce slip for subsequent processing and rub and scratch resistance in the ink on the finished product (these are normally polyethylene or polytetrafluoroethylene, PTFE).

Metal decoration inks are printed on to the metal surface which has been coated with a size or an enamel. The drying of the ink film is generally achieved by means of a 'tunnel oven'. The printed sheets are picked up in the end of the press by an endless chain of racks which transports them vertically through the tunnel oven, where they are subjected to blasted hot air at around 150°C for 12–15 minutes. Some ultraviolet cured inks are used where a set stack is used instead of the tunnel oven. The majority of inks are then overvarnished and stoved. Food containers are coated with a lacquer on the reverse side of the sheet and stoved at about 200°C to provide an internal coating on the finished can.

Two-piece cans and plastic food containers are printed by dry offset, sometimes called 'letterset'. This is an amalgam of relief and offset printing. The printing image is carried by a relief surface, normally a photopolymer plate which is inked by a train of rollers delivering a 2–4 μm film of ink to the plate. This ink film is then transferred to an offset blanket. Up to six colours are transferred consecutively to the blanket to complete the design, which is then transferred to the container to be printed in one pass of the blanket cylinder. The overvarnish is then applied by a satellite printing unit. The drying process used depends on the type and size of the container being printed.

Infrared drying is normally used for food containers whilst traditional hot air is used for other types of container. Some ultraviolet curing is used either on-line or off-line in a tunnel oven.

Dry offset printing is a high-speed process and the ink is specially formulated to meet the demand of rapid drying. Conventional oil-based oxidation polymerisation drying varnishes are unsuitable and these are replaced by solvents and plasticisers or acrylate-based ultraviolet curing vehicles.

One of the biggest problems of lithographic printing is the presence of water, and maintaining a correct balance between this water and the ink. Over the years there have been a number of experimental systems to print without water. One of the most successful is that marketed by Toray Industries of Japan. In their process, the printing plate has a photopolymer to give the printing image by the normal method of exposure to a negative or positive. The non-image area remaining after image processing is, however, not a metal but siliconised rubber which is oleophobic and will not accept an ink film. Therefore, on the press the water is eliminated; this in turn eliminates the problem of controlling the emulsification problems in ink formulation. A thicker ink film is possible, without adversely affecting dot gain in image transfer. The printed image is improved in appearance over traditional lithographic printing. However, largely due to the absence of water, a considerable amount of heat is generated during printing. This heat has an adverse effect on the performance of the ink on the press. In order to print by this method, presses have to have inbuilt cooling systems to maintain the temperature of the ink and the plate at a constant 32°C. Additionally, the plate surface in the non-image areas is more delicate than anodised aluminium and easily damaged. Despite these problems the process is gaining in popularity due to higher productivity and image quality.

2.6 The printing industry

The industry tends to be divided into specialist groups, viz:

- Publication printing to include newspapers, periodicals and books.
- Packaging printing which includes paper and board, metal, glass, plastics and wood.
- Business and consumer products such as advertising literature, business forms, calendars, diaries, greeting cards, catalogues, directories, envelopes and stationery.

Table 2.1 Printed products. (Source: PIRA, *UK Printing and Publishing Statistics*, 1994, p. 2)

Advertising literature
Atlases/maps
Books
Business forms
Calendars
Cards – greetings cards, postcards, business cards, invitations
Catalogues (including mail order), brochures (including travel), leaflets
Diaries and organisers
Direct mail
Directories
Envelopes
Financial print – annual reports, bond prospectuses, mergers and acquisitions documents, offer for sale documents, rights issues, share issue documents
Gift packaging/wrapping
Holograms
Inserts
Magazines and journals
Materials – glass, china, ceramics, plastics, textiles
Newspapers
Outdoor advertising, posters
Packaging – paper, plastic, metal, glass, wood, labels
Security print – bankers' drafts, bonds, certificates of deposit, credit cards, credits and drafts, currency, international money orders, passbooks, passports, personalised cheques, personalised credits, phonecards, postage stamps, tickets, vouchers and coupons
Stationery
Timetables

2.7 Applications of the printing processes

2.7.1 Letterpress

This process is generally on the decline in the Western world due to the relative slowness of printing surface preparation and press speeds compared to other processes. Letterpress has been unable to adapt to the use of digital image preparation of both text and illustrations at a price which allows it to compete with other processes. However, small general jobbing printers and some continuous stationery producers still use this process. The newspaper industry, which was dominated by rotary letterpress, has totally changed production to offset lithography or flexography.

2.7.2 Flexography

Flexography is a buoyant process, being used for a wide variety of printing in packaging, publication and business and consumer products, such as paperback books, newspapers and comics.

The process was originally designed for simple paper-bag printing. Due to rapid technological developments in printing machine design and printing surface preparation, flexography is taking an increasing market share in a large variety of printed products. It is ideally suited to printing on any substrate in reel form, such as films, foils, paper or board, especially for packaging and wallcoverings where it is in direct competition with gravure, particularly at the cheaper end of the market. Flexo is ideally suited to broad or open design work as it normally requires somewhat wider design tolerances than gravure. However, with increased press and plate sophistication, more high-quality work is being produced by flexography. The 'kiss' impression of the modern flexo press reduces the image distortion often associated with flexo printing at speeds of around 1500 feet per minute. The ease now possible in printing image changeover on the press has made short runs financially viable. The photopolymer plates used are capable of run lengths in excess of 300 000 impressions.

The presses used are very large and relatively expensive, depending on the type of printing they are designed to produce. A label-only press will cost around £155 000, a four-colour business forms press costs around £500 000, and high-speed wide web multicolour presses cost well in excess of £1 million.

2.7.3 Gravure

Gravure printing is classified into publication printing and package printing.

Publication printing is normally undertaken on fixed-size, custom-built presses costing many millions of pounds. The process is essentially reel-fed on supercalendered paper. The printing cylinder is normally eight A4 pages (e.g. 1.5 metres in length). The multi-unit press prints both sides of the web in four colours with in-line folding to produce finished products. The process is capable of very high quality output, at printing speeds of around 35 000 impressions per hour.

The copper printing cylinders are chromium plated before use on the press. By a process of de-chroming and re-chroming these printing cylinders, their press life can be extended almost indefinitely. An example of this is postage-stamp printing. Many millions of impressions are required of the image at the same consistently high quality and colour, as colour variations or printing errors become prized collectors' items.

Publication runs of millions of copies are normally printed by gravure, but with modern pre-press technology and press design, shorter runs of around 150 000 impressions are now financially viable.

Packaging printing by gravure is produced on standard-sized machines running at about 300 metres per minute but with a variety of cylinder widths to suit the product, from the narrow 45 cm web, to the 1 metre standard web. The cylinders are stored for reprinting; this means that there are a lot of stored spare cylinders which increases production costs. The quality of product is very high and this often offsets the costs in the long run. As has been previously stated, multimillion impressions are normal from chromed cylinders.

Specialist products are also printed by gravure such as vapour-phase transfer papers for textile printing, wallpapers, vinyls, both wall and floor coverings (where 4 metre cylinders are used), and papers for decorative plastic and wood grain laminates.

2.7.4 Screenprinting

A wide range of printing machines are available to match the wide variety of applications of this versatile printing process. At its simplest, the flatbed press with totally manual operation will cost only a few hundred pounds and have a printing speed of 50 or 60 impressions per hour per colour. Totally automatic flatbed machines with on-line drying cost in excess of £250 000 and have output speeds around 1500 impressions per hour.

Faster speeds require cylinder presses where the substrate, attached to a cylinder by vacuum, passes under a reciprocating screen with a fixed-position squeegee. This process is capable of up to 5000 impressions per hour. An adaptation of this press design, called a 'container press' is used to print three-dimensional objects such as bottles, containers and ceramics where the object to be printed replaces the position of the printing cylinder.

Rotary screen presses are where the squeegee and ink supply are fitted inside a rotary screen cylinder. This process is used for label printing on narrow web presses, and wall and floor coverings and plastics on wide web presses. This process is in direct competition with flexography and gravure in both output and costs. Specially adapted presses called 'carousel presses', with up to 12 sequential printing stations, are used to print textile garments such as sweat and T-shirts.

Screenprinting is used for point-of-sale displays, printed signs on a variety of substrates, instrument panels, textiles and posters, etc., for both long- and short-run work. However, non-impact printing technologies are taking an increasing market share of short-run work – see Chapters 4, 5 and 6.

2.7.5   Lithography

Lithography is commercially divided into sheet-fed and web-fed processes. Sheet-fed litho is the workhorse of the printing industry, utilising presses in a wide variety of sizes, printing units and speeds. At its simplest, the A4 small offset single-colour press is used for a wide variety of printed products both in the office environment and on a commercial scale for producing reports, in-house stationery, flyers and general printing of small sheet sizes. By the use of modern platemaking technology, quick changes of images are a feature of the fast turnround of the work capable of these machines. Printing speeds of up to 12 000 impressions per hour are obtained from these small versatile machines. At the other end of the scale are presses with a sheet size of 1 × 1.4 metres. These presses are normally configured as in-line multi-unit presses with in-line varnishing as an optional feature.

The modern A3 and larger multicolour presses have computer-controlled inking and damping systems. This, coupled with automatic plate positioning on the press, increases productivity by reducing make-ready times and controlling colour variations during the press run. Very short runs are financially viable with such control systems. Depending on the type and make of the press, speeds of between 10 000 and 15 000 impressions per hour are available. Multi-unit perfector presses are used for book printing in both black-and-white and in colour.

Presses can cost anywhere between £41 000 for an A3 two-colour press running at 10 000 impressions per hour to £1.6 million for a 720 × 1040 mm six-colour press running at 15 000 impressions per hour.

2.7.6   Web offset

A multi-unit web-fed offset process, with presses utilising reels of varying widths, is capable of printing four colours on each side of the web. This is ideal for publications and packaging work. A press with 4 × A4 images across the web is built to run at 35 000 impressions per hour but is normally run at around 22 000 to 25 000. The smaller 2 × A4 webs run at around 20 000 impressions per hour.

There is fierce competition between sheet-fed offset and narrow-web offset for short-run multicolour work. Large web competes directly with gravure in publication work and all three process (gravure, flexo and web offset) compete for long-run packaging and label printing. Newspapers are increasingly using web offset as their printing process where the application of digital imaging in black and white and colour allow for rapid plate preparation and high-speed printing with computer-controlled presses.

Presses are relatively expensive, being well in excess of £1.5 million, but are competitive with other processes due to high productivity coupled to high quality of output.

2.8 Commercial aspects of the British printing industry

The British printing industry is the UK's sixth largest manufacturing industry and the fourth largest export industry. There are some 18 000 companies in the industry, employing around 3 million people. The total output of the industry is valued at around £18 billion per annum. How this is spread between the various technologies is shown in Table 2.2. In order to keep abreast of the rapidly changing technologies and to maintain output and competitiveness, the industry spends around £1 billion per annum on capital.

The total value of products by printing process in 1994 represents over 10.5 million tonnes of paper and board, and 52 000 tonnes of ink. The 1994 survey of the top 500 UK printing companies by *Printweek* showed that the UK printing industry is increasing in efficiency and productivity. The average turnover in 1994 increased by 16% on the previous year in the top 500 companies and employment increased by 2.4%. The operating profits increased by 18% and the pre-tax profits by 17.5%. This reflects the upwards trends in circulation of newspapers, magazines and journals.

Table 2.2 Total value of products by printing process (1994). (Source: PIRA, *UK Printing and Publishing Statistics*, 1994, p. 15)

Lithography	45%
Gravure	12%
Flexography	22%
Letterpress	5%
Screen printing and other plate processes	8%
Electronic, ink jet and other plateless processes	8%

Book titles also increased in 1994. Cartons and packaging also showed an increase as did labels. Stationery showed a slight decrease in volume but exports rose.

2.9 Electronic imaging

Traditionally, all text matter was set in hot metal. The metal type was used directly in letterpress to produce the printed product. In gravure and lithography, the type was carefully proofed by letterpress techniques and photographed in a large process camera. The subsequent negatives or positives were used to produce printing surfaces.

In 1956, the first commercial machine was installed in Glasgow to produce text directly by photographic means. Over the next 10–15 years, a wide variety of analogue phototypesetting machines appeared on the market. The text was keyed into a machine which produced punched-paper type to drive the photosetting machine. The photosetter had a matrix of typefaces in photographic negative format. Photographic paper was exposed via an optical system to produce a paper photographic positive. This was imposed into page format and rephotographed to produce a master image for platemaking. By the 1970s, direct-entry digital typesetters were available using a raster scan laser to generate the text.

In the 1980s, the integration of text and graphics took place and the incorporation of RIPs (raster image processors) now dominates typesetting. Illustrations were produced by rephotographing original artwork to produce photographic negatives or positives which were incorporated with the text during page planning, prior to printing surface preparation. Continuous tone artwork, including photographs, were rephotographed to the size required for printing. A half-tone screen was placed in the camera; this broke up the image into a series of dots which allowed the illusion of tone to be reproduced on the printing press by the use of a single thickness of ink film. Half-tone screens are available in a variety of line rulings from around 24 lines/cm (60 lines/inch) used to print on newsprint to 80 lines/cm (200 lines/inch) for printing on very high quality coated papers (53 or 60 lines/cm being the most commonly used ruling used in litho). Half-tone dot shapes may vary to produce differing effects. The most commonly used are the conventional square dot and the chain dot, used where flesh tones dominate the illustration (Figures 2.5 and 2.6).

Colour illustrations were colour separated at the time of photographing them to the size required for printing. Exposures were made separately through a red, green and blue filter. The exposure to produce the black print was usually made through a yellow filter. This colour separation was usually done in continuous tone to allow retouching to take place. Half-tone positives were made from these negatives to provide images for the

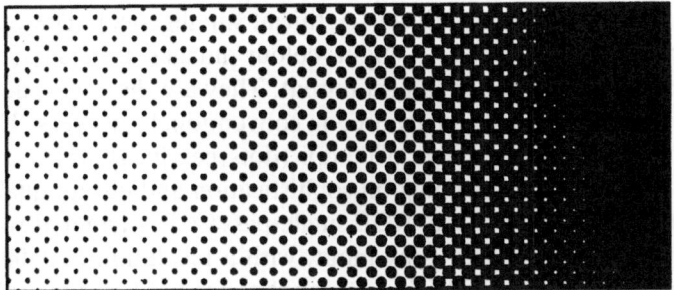

Figure 2.5 Conventional square-dot screen pattern.

Figure 2.6 Chain-dot or elliptical screen pattern.

cyan, magenta, yellow and black printing plates. The black printer was known as the key, hence CMYK printing.

2.9.1 Scanning

Currently, all artwork is scanned to size and digitised for further image processing. This technique can also be used for text printed for OCR (optical character recognition).

Scanners vary in size and architecture: the drum scanner dominated the market for 30 years. These drum scanners are used for high-quality colour separation, having high productivity and versatility of output, either film images, on-line digital information transmission or off-line digital storage on tape, disc or optical data storage (see Chapter 7). The digital information of the images is then passed on-line or off-line to a CEPS or Colour Electronic Pre-press System for further colour editorial work and preparing the final images for output. This equipment requires highly skilled staff to operate it to maximum effect. The investment required is in the order of £1 million. The current trend is towards flatbed CCD (charge coupled device) scanners with AI (Artificial Intelligence). These scanners were originally developed to input colour images to DTP (desk top publishing) systems for non-impact output. They are, however, developing very rapidly and are increasingly used in conventional pre-press. They have resolutions from 1500 to over 3000 dpi (dots per inch) with size scaling facilities. Prices vary depending on sophistication and output quality from £30 000 to around £75 000 e.g. RGB single-pass CCD scanner at A3+, both reflection and transparency copy input, with a resolution of 3800 dpi and enlargement capability to 2000% working at a speed of 12 scans per hour at a size of 305 × 457 cm, is currently marketed at £29 500. They can match drum scanners on quality but at the moment not on output.

Most scanners can be operated from a Mac, PC or workstation platform. By the use of the software in the driver platform the use of a CEPS is eliminated. The platform can perform image processing by the use of a package such as Photoshop, and layout programmes by the use of Quark X press or similar package. Professional Photo-CD from Kodak (see Section 7.8) will scan originals of 100 × 125 mm (4 × 5 inches) at a resolution of 4096 × 6144 pixels per colour, stored on a CD for input to a front-end system.

The use of digital cameras working up to 7000 × 5000 pixels eliminates the need for scanners completely.

2.9.2 Output

The output from a digital pre-press system can take a wide variety of formats. At its simplest, the images can be displayed on a monitor for 'soft proofing' by the client. Hard-copy proofs can be produced by one of the non-impact printing technologies such as ink jet (see Chapter 5), thermal transfer (see Chapter 6) or electrophotography (see Chapter 4). The digital information can be used to drive a laser platemaker, a laser cutter for flexo, or an electromechanical engraver for gravure. More commonly, at the moment, the information is used to drive an imagesetter to produce films for platemaking. However, the most recent use for digital pre-press is to use the information directly into a digital press.

2.9.3 Proofing

The industry standard for pre-press contract proofs between client and printer is the Cromalin by Du Pont or the Matchprint by the 3M Company. Both these systems require the digital information to be output to film before a hard copy is possible. This is expensive in terms of time and materials, and, if there are any alterations, the whole outputting process has to be repeated.

'Soft' proofing is being undertaken more frequently now, where the digital files are transmitted to the client for viewing on the VDU system. This can only realistically be used for positional proofing. However, it is possible for the client to have a non-impact printing technology driven proofing system. The ISDN (Integrated services digital network) system can be used to carry large amounts of information to drive both soft- and hard-proof systems. Ink jet using Hertz technology in the system by Iris Graphics (see Chapter 5) has proved to be capable of excellent pre-press colour proofs, as has dye sublimation thermal transfer shown by the Kodak 'Approval' system (see Chapter 6). Electrophotography, using both laser and LED (light-emitting diode) arrays as exposing devices, has proved capable of giving excellent digital proofs.

Both Cromalin and Matchprint are available in digital format. Cromalin uses ink jet technology and Matchprint is electrophotographic. One of the problems associated with digital proofing is the lack of the ability to produce a proof in the half-tone dot pattern used for conventional litho printing. However, the introduction recently of frequency-modulated screening (FMS) by a number of electronic pre-press equipment manufacturers may well negate this problem. The output control algorithm of a non-impact printer is an ordered dither of the spot placement. In FMS it is a random dither, making the output from the two systems more compatible in visual representation of colour. The laser spot placement of a standard output from an image setter to produce films for proofing and platemaking is controlled by a half-tone look-up table and the half-tone dots are composed of a cluster of recorder spots which can be controlled to give a variety of screen rulings in lines per centimetre and a variety of dot shapes; this is area-modulated screening (AMS). The same dot-placement technique is used in conventional imagesetters to produce the letter forms required for text output, which are produced as films with text and illustrations integrated to form the required page layouts. Figure 2.7 shows how FMS produces tones. The illustrations are of the Agfa Cristal Raster process. FMS would also appear to make digital imaging direct to plate more attractive in terms of speeds of production and quality of output.

There is no doubt that FMS eliminates moire patterns and rosettes, giving colour printing a smoother, more continuous tone appearance. It also gives better definition and better depth of colour than conventional AMS techniques.

2.9.4   Digital platemaking

The production of films for platemaking for text, line and half-tone illustrations directly from digital information is now standard practice in the printing industry. However, computer to plate (CtP) or digital platemaking is, as yet, not very common. There were only two systems commercially available in 1994 with around forty installations worldwide.

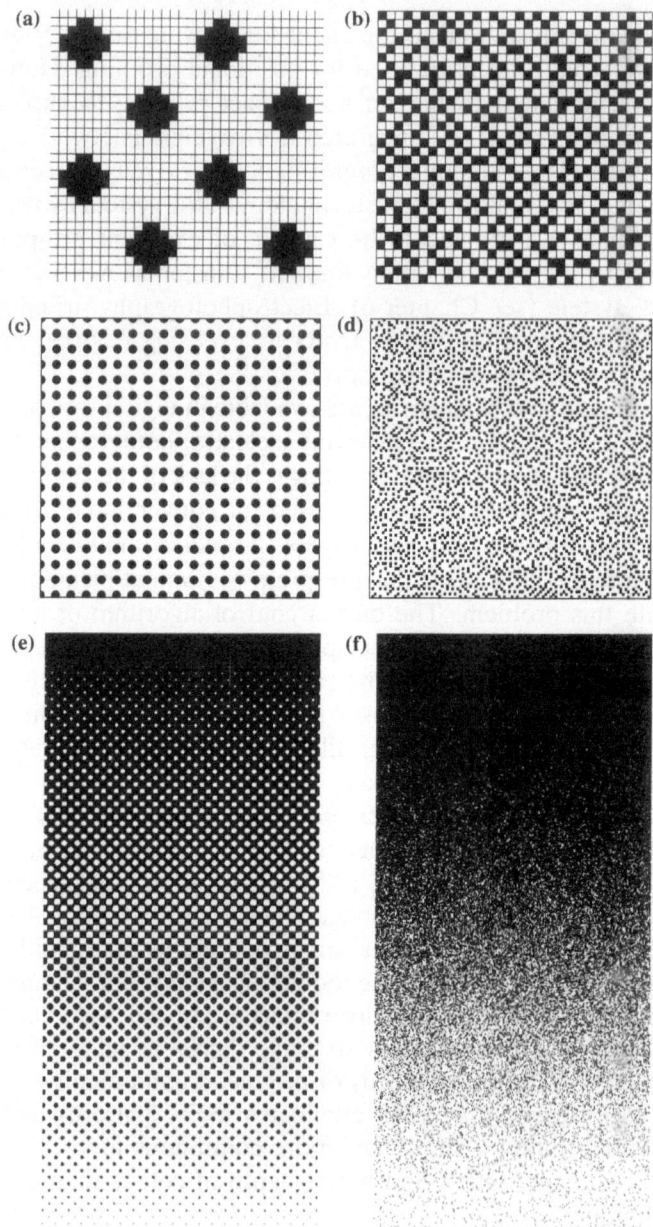

Figure 2.7 Agfa Cristal Raster, an example of frequency-modulated or stochastic screening. (a) Half-tone dots are made of clustered recorder spots. (b) CristalRaster microdot placement does not follow any structure, but is randomised. (c) 200 Ipi 25% tint at 20× magnification. Conventional screening deconstructs an image with screen grids. (d) 25% tint at 20× magnification. CristalRaster preserves all of the image data for photorealistic images. (e) Conventional screening vignette shows gray levels represented by equally spaced half-tone dots. The range of gray levels is limited by the resolution of the output recorder. (f) CristalRastar vignette is made of microdots whose distribution varies according to tone value.

Both of the currently available systems are based on high-contrast silver halides for imaging by laser exposure.

The advances in RIP technology with advanced software to deal with spreads and chokes and colour management, coupled with 32-array lasers used on an external drum, will help the development of CtP technology. There are six other CtP systems due on the market in the very near future; the plates are photopolymer, exposed to argon ion lasers in the majority of these systems. Frequency doubled YAG lasers are predicted to be used increasingly in this developing field. However, there is one application of digital information to printing surface preparation which was shown at the 1993 Screen Printing Exhibition. This was Screenjet by Gerber Scientific Products, USA, an application of ink jet to screen printing. The artwork is prepared in the normal way on computer and saved on disc. The disc is loaded into the Screenjet which has an emulsion-coated screen frame fitted in a registering device. The screen is imaged by an ink jet head. The ink jet ink is of a density which prevents light being transmitted. After 'imaging' the screen is given an exposure to ultraviolet light and processed in the normal way, which removes the ink from the screen with the unexposed polymer. Any number of printing colours can be imaged in this way as each screen frame is preregistered in the Screenjet machine. Image areas up to 510×737 mm at 300 dpi are available. This system eliminates the use of film entirely, gives accurate registration and reduces press set-up times.

2.9.5 Digital printing

The first direct digital printing press, the Heidelberg GT0-DI is based on a conventional litho printing press. The uniqueness of this press lies in the platemaking technology. The printing plates are exposed directly on the press from digital information. The four or five printing units of the press have their waterless direct imaging (DI) plates exposed simultaneously and registered on the press. Each printing plate unit has a 16 laser array connected to a RIP via the DI computer (which also calculates the ink flow to each printing unit). The laser resolution can be varied from 1016 to 2540 dpi giving half-tone screen ruling of up to 175 line/inch. The press runs at speeds of up to 8000 impressions per hour on a variety of paper stocks from 30 g/m^2 to 400 g/m^2 in sizes from 105×180 mm to 336×503 mm. The manufacturers claim a press make-ready time of only 15 minutes, which includes plate exposure time. This, coupled to make-ready wastage of only 25 sheets, makes this press a very attractive commercial proposition for short-run, quick turnround colour work of the high quality associated with offset litho.

The first truly digital colour printing system is the Indigo E Print 1000 –

see Chapter 4. The machine is designed for short-run colour printing on demand. It is a sheet-fed press with a wide variety of input options, from CEPS, tape, disc, DTP systems and Postscript. The press prints up to six colours duplex by electrophotographic technology using liquid toners. By the application of the offset process, printing on a wide variety of stocks is possible, from uncoated or coated paper, to card. The maximum printed sheet size is A3, at a press speed of 2000 full colour or 8000 monochrome A4, each two up, per hour, at a resolution of 800 dpi. The standard configuration is four colours but an option to print six colours is available.

As this is a true digital system, images can be changed from page to page with no reduction in printing speed, thus enabling entire publications to be printed, automatically duplexed (perfected) and with each page in the correct sequence. The output can be fed automatically into an on-line booklet maker. An additional option is the personalisation of individual copies. The advantages of this system are there are no films, no plates and no proofs (on-demand single prints replace proofs). There is no press make-ready; perfect register is automatic as is colour balance. There is no need for an expert professional operator as the quality is electronically assigned without dependence on the skills of the operator.

The Xeikon DCP-I print engine is a web-fed electrophotographic printer using a 307.6 mm wide LED array as the exposing source giving 64 grey levels at a resolution of 600 dpi on each of eight printing units, with the toners fused in a non-contact unit – see Chapter 4. This allows four colour printing on each side of the web in a single pass, at up to A3 size with full bleed. The input can be from any standard pre-press platform as the equipment is fully compatible with Postscript. The machine can cope with a wide range of paper grades and weights, the programmable cutter providing for flexible output formats either A3 or A4. The running speed at an A4 format is 4200 impressions per hour, duplex, giving an output speed for a four-page A4 colour brochure (from an A3 folded sheet) of 1050 per hour.

There are a number of digital printers now available on the market utilising different non-impact printing technologies. Ink jet has been used for a number of years in the conventional printing industry for coding, date marking and addressing. This is advancing rapidly with increasing speeds and addressability. The quality and speeds of wide ink jet systems available for the production of display materials is improving almost weekly. Typically this is drop on-demand technology, in four colours' printing on a wide variety of substrates, papers, plastics and films – see Chapter 5. Web widths vary, usually from 1 to 5 metres. These machines are rivalling screen printing for very short-run multicolour work.

Electrophotography is used by a number of digital printer manufacturers besides the three already mentioned. The Xerox Corporation have a number of machines available, for example, the Xerox 5773 shown at Ipex

'93 running at 7.5 A4 pages per minute at a resolution of 400 dpi in four colours. Kodak and Canon also market colour printers. The Riso digital printer-copier is also capable of colour printing. Wide-format electrostatic plotters (see Chapter 8) are increasingly being used by conventional printers for large pre-press position proofs, map printing and advertising, both in monochrome and full colour, typically working at 200–400 dpi with output speeds dependent on substrate width, e.g. 54 inches wide at 200 dpi outputs at 3 inches per second (76.2 mm/s).

Magnetic printing (see Chapter 8) has been around a number of years in small format. The Nipson digital printer was launched in 1993. A web-fed digital printer, the Vary press works at a speed of 345 feet/minute on a web 20.5 inches wide, giving an output of 700 A4 pages (side by side) per minute at 240 dpi. A duplex version exists with two print engines giving an output of 1400 pages per minute, printing on papers from 40 to 240 gsm, and with flash fusing also on plastics films.

Certainly non-impact printing technologies are now integrated into traditional printing technologies making separation an irrelevance in the future.

2.10 Printing and the environment

There is a significant amount of UK and European legislation to control environmental pollution from manufacturing industries of which printing is a part.

Atmospheric pollution is an area where printing has been criticised due to solvent vapours. It should be borne in mind, however, that the printing industry as a whole emits only about 2% of the amount of gaseous emissions that come from motor vehicles. The main problem of VOCs (volatile organic compounds) comes from ink solvents, press-room wash-up solutions and alcohols in fount solutions. Water-based inks are being used increasingly as are ultraviolet curing inks, both of which are free from volatile solvents. New types of solvent cleaners are being introduced for press cleaning. Vegetable cleaning agents are replacing petroleum distillates to clean rollers and blankets. Blends of water and slower evaporating distillates are widely used in the USA and are gaining in popularity in Europe.

Alcohols are being replaced in fount solutions by quinones and other non-volatile substances. In large flexo and gravure printing houses, solvent recovery and recycling is now common practice. The increase in the use of water-based products has highlighted the need for waste-water treatment. This includes the removal of heavy metals and colorants from ink wastes and saturated wiper cloths.

Chemical waste disposal from silver halide processing is very well controlled in the printing industry and causes no concern for environmental pollution.

Waste disposal in the UK is controlled by the Environmental Protection Act Part II 1990 and water pollution is subject to both the Water Resources Act 1991 and the Water Industry Act 1991. The basic principles of these Acts are that the 'polluter pays' and introduce a 'duty of care', i.e. the producer of waste is responsible for its proper disposal.

The environment and its protection is taken very seriously by the British printing industry.

References

Gregory, P. (1994) Modern reprographics. *Rev. Progress in Coloration*, 24, 1.

Further reading

British Printing Industries Federation (1992) *Introduction to Printing Technology* (4th edn).
Williams, C.H. (1992) *The Printers Ink Handbook*, Maclean Hunter Ltd.
PIRA International (1994) *UK Printing and Publishing Industry Statistics*.

3   Silver halide photography
P. BERGTHALLER

3.1   History and evolution of photographic imaging systems

A short introduction to the history of silver halide photography from early portraiting and documentary imaging to the latest development of Photo CD and digital printing systems may be justified by the need to demonstrate the major challenges, problems and ideas in classical silver halide photography and to present new options for the electronic age.

3.1.1   Photography in black-and-white

Many problems in photography are of historical interest only: in black and white the search for permanence of images, greater convenience, superior definition and higher photographic speed has been discontinued. Other topics such as stability to process deviations have become more important. Estimates of the ultimate limit in photographic speed concentrate on 3200 ASA. Silver halide photography in black-and-white has reached a state of maturity where the need for invention hardly exists any more.

After 1945, colour photography displaced black-and-white photography as the dominating photographic medium, and during the past twenty years key inventions of the space age, e.g. high-definition scanning and digital data transfer, have penetrated into photographic imaging. Still video-camera systems have become established at the lower end of the market.

Imaging technologies of military origin have been implanted into civil life, e.g. multispectral photography for satellite-based vegetation control. The impact of digital imaging and image processing is transforming photography into an innovative medium which has to compete with many other systems.

Wherever real time images are needed or the genuine qualities of the silver halide systems are not important, silver halide photography is on the retreat. The low-quality field has largely been taken over by other imaging technologies. Examples of this are the decline of Super-8 cinefilm and silver salt diffusion technology for office copying.

3.1.2   Concepts of photography in colour: The steps of the pioneers

The present state of knowledge of all theoretical aspects of colour photography is presented in an exhaustive discussion by Krause (1989).

The discovery of colour photography is attributed to J.C. Maxwell (1858–62) who – to illustrate the Young–Helmholtz principle of colour vision – demonstrated the photographic reproduction of body colours on a projection screen by colour separation and additive mixing of three primary light colours: blue, green and red.

The next step was taken by Ducos du Hauron (1869), whose intention to reproduce the colours of an object in a photographic print led to the discovery that yellow, purple (magenta) and cyan as primary colours could be mixed in a three-colour print, a concept that became known as 'subtractive colour mixing'.

The only tool missing for the evolution of colour photography was a suitable way of extending the light sensitivity of silver halides from blue to the low-energy region of the visible spectrum and this was achieved by the discovery by Vogel of spectral sensitization (1873).

Generally, the colour information about an object is registered by three photographic colour separations and recombined into a colour photograph by transforming the silver images of the colour separations into dye images. The evolution of colour photography has been discussed by Hunt (1988).

To avoid the use of three separations for one picture, highly sophisticated manufacturing procedures for additive screens have been developed (Land, 1977). Additive colour mixing in photographic materials has been used commercially since about 1900. Possibly the last additive line screen film, Polachome CS instant reversal film, was launched on the market by Polaroid in 1982.

The implementation of subtractive colour mixing into true colour films was achieved after 1935 by three different methods which today form the basic methods of colour photography:

- removing a dye from a layer (chromolytic method; Schinzel, 1905);
- generating a dye in a layer (the chromogenic method; Fischer and Siegrist, 1914);
- image-wise transfer of dye from one layer into another by diffusion (Rogers, 1961; basis of Polaroid instant colour photography).

3.1.3   Silver halide photography today: the market

A basic impression of the silver halide photography market is provided by evaluating recent figures on its main products. Colour photography constitutes about 80% of the worldwide turnover (Table 3.1).

Table 3.1 Worldwide consumption (millions of square metres or units). (Source: *Photofinishing News 1993*)

Material	1984	1988	1990	1992
Colour negative (CN) paper (m²)	591	825	937	1012
Black-and-white (B&W) paper (m²)	127*	340	325	248
Conventional film (35-mm, rolls, discs) (units)	1700	2108	2311	2307
Photographic chemicals†	–	588	655	733

*Represents only continuous tone paper.
†Value in $ millions; the figures represent higher prices plus small volume increase.

The breakdown for 1992 shows 85.6% colour negative (CN), 8.6% colour reversal (CR) and 5.8% black-and-white (B&W) film (including 70 million of Japanese manufactured single use/film with lens cameras as the fastest growing segment).

3.2 The physical basis of silver halide photography

3.2.1 A brief look at the elementary process in the silver halide crystal

Photographic sensitivity has been observed on microcrystals of silver chloride since the late 18th century. As a phenomenon, photographic sensitivity is not restricted to silver chloride, silver bromide and silver iodide, but there is no other system where it can be generated, handled and investigated with comparable comfort and efficiency. Thus the silver halogenides became the basis of modern photographic systems.

Photoconduction and photographic sensitivity of silver halogenides are well understood. Silver halides are sensitive to actinic light and the absorption of a single photon generates a conduction band electron and a positive hole of minor mobility. Once charge separation can be maintained over a short period and over a series of single events, the exposure leads to a latent image. A series of four hits can be sufficient to make a silver halide grain developable (James, 1986).

If a silver halide microcrystal contains a small number of clusters of just a few silver atoms or electron-donating silver sulphide specks on its surface, holes can be trapped irreversibly and the photographic process can take place with a high level of efficiency. Reversible photoelectron trapping on the other hand leads to an extension of the period during which the silver halide microcrystal remains vulnerable after the first photon event. An advanced discussion of the present theoretical position is given by Mitchell (1993).

3.2.2 *The photographic emulsion* (Berry, 1977; Lapp, 1992)

The term 'photographic emulsion' relates to a dispersion of silver halide microcrystals prepared in an aqueous solution of gelatin or another natural or synthetic protective colloid. The main functions of the colloid in emulsion preparation can be seen in providing the desired growth control and preventing the crystal nuclei from undesired multiple agglomeration.

Gelatin is a polypeptide and in its 'natural' state dissolves in water above 30°C. As the binder of the coated photographic layer, gelatin has to provide the desired degree of swelling and solution uptake during processing. This has to be achieved by establishing crosslinks in the peptide structure of the gelatin binder during or after the coating process of the photographic material. This process is named 'hardening of gelatine'. Hardened gelatin swells in water of processing chemicals but it does not melt.

The silver halide microcrystals of photographic emulsions are prepared either by precipitation from a soluble silver salt and a soluble halide or halide mixture or by growth from small seed crystals and dissolving sacrificial micrate, i.e. from a preformed silver halide emulsion containing very small crystals of the same or different halide composition.

Careful control of the growth conditions produces a well-defined crystal size, size distribution and shape. In particular, the concentrations of the reactants and their variation during precipitation, the presence of growth-accelerating or growth-modifying additives and temperature control are used to enhance crystal growth along selected lattice planes and to inhibit the growth of others. Cubes, octahedra and twinned tabular crystals or sometimes also compact crystals are the preferred forms of photographically active silver halide 'grains'.

Spectral sensitization by dyes has been discussed either in terms of energy transfer or in terms of direct electron transfer from the excited dye to the silver halide microcrystal (Gilman, 1977; Tani *et al.*, 1968) but experimental evidence is now favouring the more complex view of a 'hole injection' mechanism, where the key role is played by a dye radical anion, which in a final thermally controlled step transfers an electron into the conduction band of the crystal (Siegel *et al.*, 1990).

3.2.3 *Photographic development* (Lee and Brown, 1977; Matejec, 1992)

Once a silver halide crystal bears a latent image, contact with an alkaline solution of an electron-donor compound of appropriate reducing activity leads to rapidly increasing growth of the latent image silver speck. Initially this process is controlled by silver-ion mobility within the crystal and by the limited surface area and conductivity of the silver electrode. It is called 'chemical development', in contrast to 'physical development' which can dominate in the later stages of development. In physical development, the growth of silver aggregates is maintained essentially by the transport of silver complexes within the solution. Chemical development is characterized by filamentary growth of the metallic silver, while physical development leads to compact silver of a more globular shape.

Irrespective of the mechanism, there is one condition for efficient development: neither the transport of halide ions from the surface of the crystal into solution nor the passage of electrons from the adsorbed developer moieties into the silver electrode must be inhibited.

Moderate inhibition by adsorption of developer oxidation product, bromide or iodide ions or other inhibiting anions on the growing silver electrode is an effective tool for controlling the quality of a photographic image.

Silver halides are metastable in contact with reducing agents and it is only a question of time before each silver halide crystal of an emulsion, be it exposed or not, is developed by the spontaneous formation of silver nuclei. Provisions against this kind of 'fogging' must be taken.

3.2.4 Detail rendition

The photographic emulsion is a superior medium for the capture of a low-intensity picture as it is projected by a lens. However, its efficiency is limited by inherent disadvantages of the photographic layer.

Firstly, the silver halide crystals of a layer are deposited at random and the efficiency of recording incident photons is subject to statistical control. Secondly, due to large differences in refraction between silver halide and gelatin, the silver halide layer shows strong light scattering. The resulting loss of acutance may be minimized by the use of screening dyes.

An additional loss of local information is caused by non-uniform distribution of developed silver with minimal fluctuations of density. Thus, the photographic silver halide layer is far from perfect as a signal recording medium. All attempts to establish a regular monolayer distribution of flat silver halide grains of different sensitization in different emulsion layers have led to unsolved problems. In colour development, the diffusion of oxidized colour developer from the emulsion grain to the dispersed coupler also contributes to loss of local information.

3.2.5 Development and colour: subtractive colour photography vs. additive colour photography

To the surprise of many experts, additive colour photography has survived the powerful growth of subtractive colour photography that began in 1935.

Additive colour reversal films for instant processing are being produced only by Polaroid. In an additive colour reversal instant film (Liggero et al.,

1984) a single panchromatic silver halide layer is bound to a trichromatic screen in permanent register. The silver halide layer is exposed through the screen and film base and subjected to black-and-white 'diffusion-reversal processing'. The negative is stripped off, leaving a positive on the base and the reconstruction of the colour image is done by projection through the intact screen with white light: since each coloured spot of the picture absorbs two-thirds of the projection light, one stop must be sacrificed to obtain the colour information.

It is easy to appreciate that additive colour mixing is inappropriate for colour printing. Surprisingly, additive colour photography in its original version as designed by Maxwell has proved the most reliable procedure for taking colour pictures of astronomical objects without any reduction in quality compared to astronomic black-and-white photographs (Malin, 1993). The long duration of the procedure, where blue, green and red separations are generated in three sequential shots on hydrogen-sensitized panchromatic black-and-white photographic plates, is commonly used in astronomy. After development, the separations are combined by printing in succession on colour film, which is later used as the master for final prints.

It must be kept in mind, however, that in subtractive colour photography, the colour separation step is done according to additive principles: coloured light as a mixture of blue, green and red constituents in different amounts is recorded by three detector elements. Subtractive mixing of colours results from the processing step where the subtractive primary colours yellow, magenta and cyan are either generated or removed from the layer arrangement. Subtractive colour mixing is understood as a superposition of a yellow, a magenta and a cyan filter layer which combine to neutral black at maximum density (Figure 3.1). In a transparency the

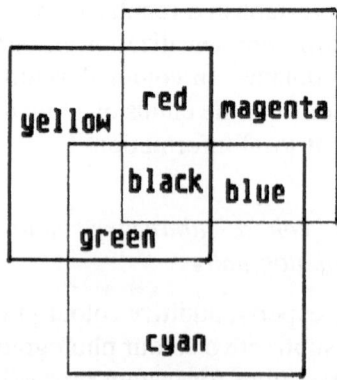

Figure 3.1 Subtractive colour mixing by superposition of filters.

appropriate optical density of each layer is 3 or higher to provide a genuine black while the preferred density values of a colour print are about 2.

3.3 Photographic materials

Film and paper form the core of the photographic market. Their volume determines the turnover of the photofinishing industry. Recent figures show that the demand for film and paper will rise further in the near future and consequently the photofinishing business will also grow. Fast-growing nations tend to bypass the black-and-white era.

The average European customer buys roughly two colour negative (CN) films per year, takes most pictures during holidays, empties the camera within 8 weeks and orders about 50 prints. This kind of business forms the bulk of the photographic market.

3.3.1 Film

All photographic materials coated on a transparent base are called 'films'. Films are produced in different formats according to their use and camera format standards. Films for amateur use differ considerably from those for professional use as regards their overall properties.

3.3.1.1 Black-and-white film. This field of business has been declining for more than ten years. The black-and-white market comprises half-tone film, line film, radiographic materials (e.g. X-ray film) and materials based on the silver salt diffusion process (e.g. phototyping film).

3.3.1.2 Colour film. The layer arrangement of all modern colour films is basically the same (Figure 3.2). Within the last two decades, the average number of layers has increased and today at least two layers are combined to a colour pack, with the low-speed layer near the film base. The coating is carried out by modern multiple coating equipment to keep production rates high and costs down.

Professional colour films are produced as daylight films and also as films for exposure in artificial light. Daylight films give optimum results on exposure to daylight with a colour temperature of 5500K, while films for artificial light are designed for exposure to light of 3200K.

The bottom coat next to the substrate is usually an antihalation layer containing black colloidal silver. Since metallic silver is bleached in the course of processing the antihalation layer is decoloured in the finished film.

The low-speed red sensitive layer is applied next to the antihalation layer. A medium speed layer can follow. Normally, the high-speed red

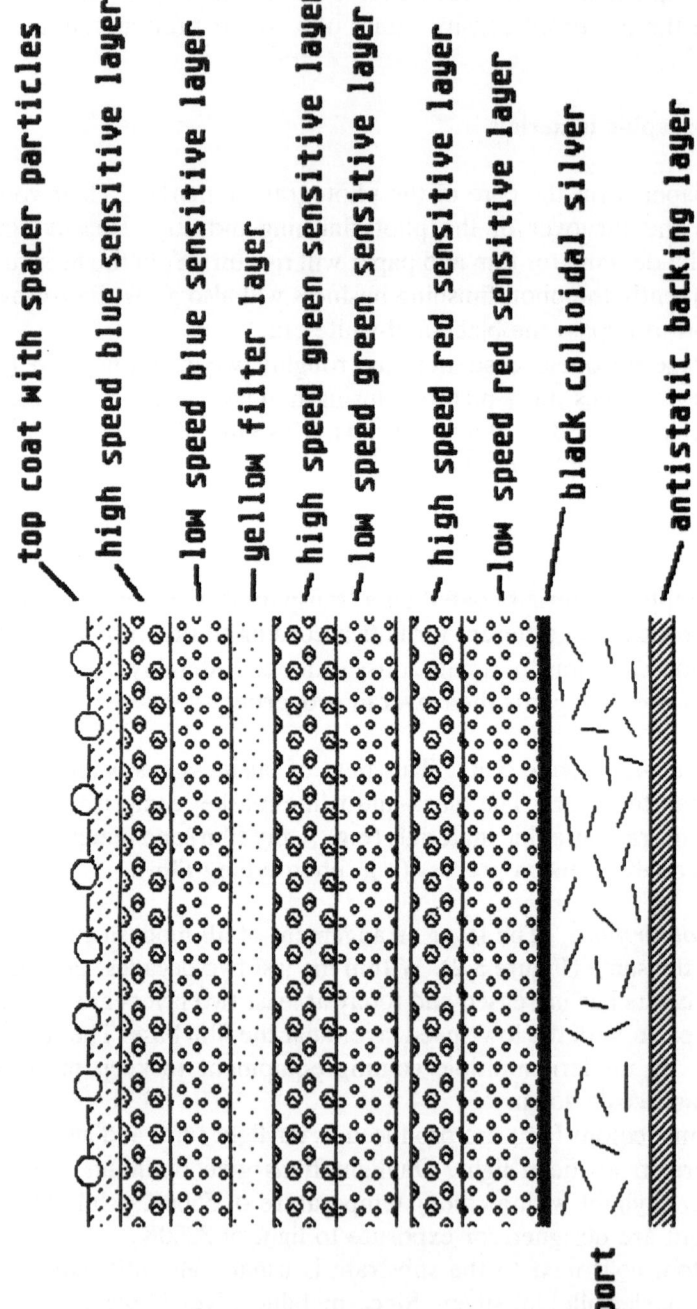

top coat with spacer particles

high speed blue sensitive layer

low speed blue sensitive layer

yellow filter layer

high speed green sensitive layer

low speed green sensitive layer

high speed red sensitive layer

low speed red sensitive layer

black colloidal silver

antistatic backing layer

support

Figure 3.2 Colour film: layer arrangement. Simplified. Not to scale.

sensitive layer containing coarse grain emulsions is applied next, but in certain cases where speed is the most important demand and acutance is less critical, the low-speed red sensitive layer and the low-speed green sensitive layer are coated first, and the high-speed green sensitive layer and the high-speed red sensitive layer are combined to a separate pack. The blue sensitive layers are coated in the highest position.

Intermediate layers containing a scavenger for oxidation products of the developer are applied between emulsion layers of different colour response. A red filter layer may separate the red sensitive pack from the green sensitive pack to improve colour separation and acutance. A yellow filter layer prevents blue light from passing into the green and red sensitized layers.

The top coat may be applied as a stack of single layers containing additives such as spacer particles which have to provide smooth transport in the camera. It may also contain the hardener, a plasticizer and a balanced mixture of filter dyes providing the desired colour balance.

3.3.1.2.1 Colour negative (CN) film. Colour negative film (Figure 3.3) usually has the ability to master high contrasts and overexposure up to three stops without a visible loss of quality. Prints from a seriously overexposed medium-speed colour film show reduced acutance and colour separation and may be slightly out of colour balance. Due to the use of DIR (development inhibitor releasing) couplers the low granularity is

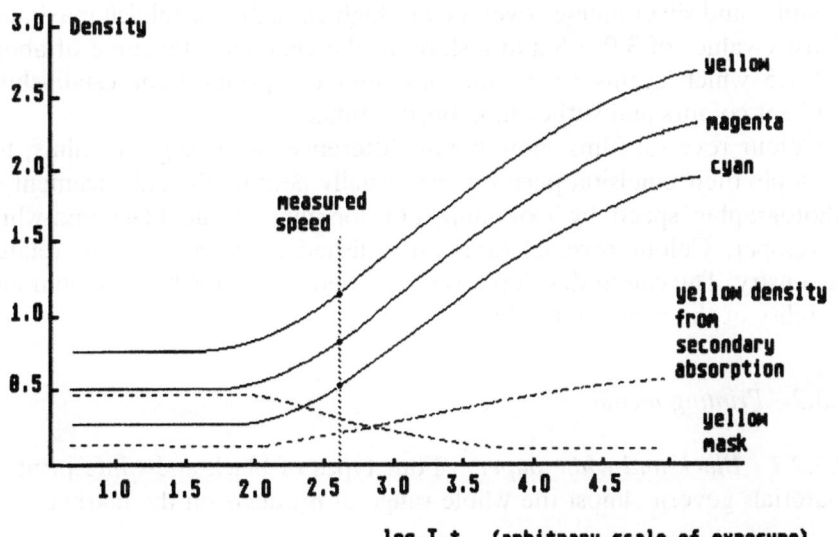

Figure 3.3 Characteristic curves of a colour negative (CN) film; slope 0.7.

maintained on overexposure while on underexposure, granularity and reduced colour saturation become visible in the print.

As a prerequiste for good colour reproduction, the single-colour gradations in yellow, magenta and cyan must be tuned to provide neutral grey in the print from the threshold to maximum density. Chromogenic dyes exhibit considerable secondary absorption which must be compensated either by reduction of, for example, 'true yellow' with the consequence of reduced chromaticity, or by masking, for example of the yellow densities of magenta and cyan. The masking problem is discussed in detail by Hanson (1950). Besides the complementary colours and inverted density values, it is the staining effect of the mask couplers which gives a colour negative its typical appearance.

CN films are finished in the C-41 process (the common name is Flexicolor) or comparable processes of other companies. For activity and granularity reasons, the colour developer is CD-4. The usual silver content of a colour negative film is 10–15 g/m^2 and the thickness of the coatings is 22–30 μm.

3.3.1.2.2 Colour reversal (CR) film. Colour reversal films (*Colour Section:* Figure 3.4) are different from colour negative films in that they are unmasked and do not contain DIR couplers. The use of DIR couplers would not make sense because CR films undergo a black-and-white first development as the quality determining step where DIR couplers are ineffective. Occasionally, non-colour forming DIR compounds are used.

A reversal exposure or fogging treatment renders the residual undeveloped silver halide developable in the second or colour development. Coupler and silver halide coverage are high enough to establish maximum density values of 3.0–3.5 and a slope of the characteristic curve of about 1.3–1.5 which seems to be the optimum compromise for establishing brilliant colours and sufficient exposure range.

Colour reversal films show minor differences from negative films, for example their emulsion performance usually permits the enhancement of photographic speed by extending development in the black-and-white developer. Colour reversal films are finished in process E-6 or related chemistry. The colour developer is CD-3, notable for the better colour and stability of the azomethine dyes.

3.3.2 Printing media

3.3.2.1 Black-and-white paper. Four types of black-and-white printing materials govern almost the whole range of products on the market:

- photographic papers on baryta-coated paper base;
- black-and-white paper on resin-coated base;

- dry silver printing materials; and
- instant printing materials based on silver salt diffusion (such as line papers or continuous tone papers).

Black-and-white photography is the domain of business photography as well as that of professional photography or artistic photography. Amateur black-and-white photography may be regarded as a vanishing genre. Black-and-white prints on fibre based baryta paper have never lost their high reputation as fine art prints. Baryta paper is the medium of choice for obtaining black-and-white prints of outstanding long-time stability.

Resin-coated black-and-white paper is produced in fixed gradations and as multigrade paper where the gradation of the print is determined by exposure filters. The reputation of resin-coated black-and-white papers has suffered to some degree by the tendency of polyethylene laminates to become brittle and lose adhesion to the paper base when the prints are exposed to sunlight over a long period. On the other hand, resin-coated papers show convincing advantages in handling. In particular, the low uptake of processing chemicals and the short rinsing and drying time make them a favoured material for professional use (Parsons *et al.*, 1979).

Thermally processed printing materials like 3M's well-known 'Dry Silver' material are based on soap-like silver salts of organic acids (e.g. silver behenate) and contain only a little silver halide as a sensitizer. They are sold on the market in many grades and for many applications, including films for aerial photography, cathode-ray tube imaging, laser imaging and for micro-film enlargements where they compete with xerographic procedures (see Chapter 4).

A convincing advantage of all thermal printing materials is the simple and entirely dry processing step which comprises only a passage of the material through a heating device at temperatures above 100°C (see Chapter 6). The material contains a toner–stabilizer mixture to stabilize it against post-process density increase. The remaining printout tendency limits the use of Dry Silver paper to applications where true long-term stability is not a basic demand.

3.3.2.2 Colour paper and colour prints. Colour prints can be made from colour negatives as well as from colour transparencies. In the first case, a series of pictures is taken on CN film which is developed to give a strip of colour negatives showing inverted density values and complementary colours. The areas of minimum density are stained by the uniform reddish coloration of the unreacted 'mask couplers' as a characteristic of the type of film used.

The information content of a colour negative is largely inaccessible to the human eye and – apart from evidence of underexposure – even experienced people in photofinishing laboratories are not able to really predict the quality of a paper print from the appearance of a negative. Since the introduction in industrial photofinishing of modern high-speed printers with total film scanning this is no longer a problem.

In the second case, reversal colour prints are made from colour transparencies – generally framed – and normally only from a small series of slides of exactly the same colour balance. Deviations of colour balance in a series of colour transparencies can, however, usually be recognized by experienced operators.

Predetermining the quality of a print from a colour slide seems easier at first sight but this does not make the reversal print a superior product. The reason for this lies in the fact that colour slides are unmasked and also that slides show higher contrasts than colour negatives. Colour reversal paper has to compensate for both drawbacks, the secondary absorptions of the image dyes as well as the high contrast of the slide, within the technically available limits. Prints from slides are of minor importance on the market.

3.3.2.2.1 Colour negative (CN) paper. Colour negative paper accounts for the major production volume in the photographic industry. A reliable figure for the total volume of colour negative paper produced in 1994 is 1.1 thousand million square metres.

Designed for the needs of the photofinishing industry, CN paper has grown into the status of a cheap article of high and reliable quality. Over the past twenty years, the production volume has increased to five times that of 1974 while over the same period, sales prices have dropped to about 40%. At present, the production volume in black-and-white paper is estimated at slightly below 10% of that of colour paper.

Since the introduction of colour papers for the Rapid Access process by Eastman Kodak in 1986, the market share of colour paper based on silver chloride has gone up worldwide from about 5% in 1987 to almost 90% in 1994.

Colour negative paper based on silver bromide may possibly be withdrawn from the market in the future. The switch from EP-2 (bromide paper process) to RA-4 was probably completed by 1993.

For economic reasons, colour negative paper contains only three colour-forming emulsion layers. Silver chloride emulsions have no intrinsic blue sensitivity and there is no reason to protect the green and red sensitized layer from blue light by using a yellow filter layer. An essential feature is the specific sequence of the layers: the blue sensitized (yellow-forming) layer is applied next to the base with the green-sensitized layer (magenta) following it and the red-sensitized layer (cyan) at the top. Although cyan dyes absorb ultraviolet light and therefore protect the magenta and yellow dyes from light to some degree, the main dye-protecting function comes from an ultraviolet-absorbing layer containing a 2-hydroxphenylbenzo-triazole derivative.

Display films are different from CN paper in that they contain higher amounts of couplers and silver halide to provide the higher optical density needed for displaying. Display films also need a sandwich arrangement of UV-protecting layers to protect the dyes from incident light on both sides.

The absorption curves of azomethine dyes in CN paper are shown in Figure 3.5.

3.3.2.2.2 Colour reversal (CR) paper. Colour reversal paper differs from colour negative paper in many ways. Firstly, it is a bromide paper; there is no colour reversal paper based on silver chloride emulsions on the market. Secondly, because of the high intrinsic blue sensitivity of the silver bromide emulsions, colour reversal paper contains the blue-sensitive layer at the top and a yellow filter layer of colloidal silver below it. Furthermore, it is processed in a reversal process with a black-and-white development as the first step.

From an economic point of view it does not make sense to produce two types of colour paper for two different processes, but none of the attempts to produce colour reversal papers based on direct positive emulsions and designed for the negative process has overcome the inherent difficulties.

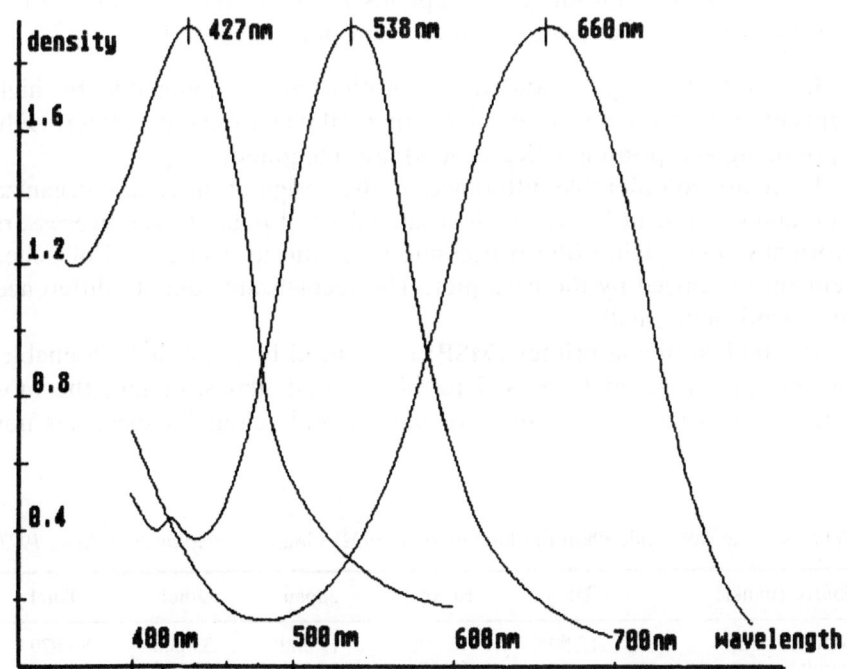

Figure 3.5 Absorption curves of azomethine dyes in colour negative (CN) paper.

Electronic reversal by the CRT printer of Agfa's Digital Print System introduced in 1991 has given almost ideal performance in producing prints from colour slides by means of CN paper and RA-4 chemistry (Heinen, 1988). This helps a finishing laboratory to run only one paper process.

Electronic reversal is only one of the many possibilities for digital image processing and is also the best means of improving the saturation of colours and compensating the higher contrasts encountered in slides. Pictures from old slides or from negatives with partially faded dyes can be produced by digital printing in otherwise unattainable quality.

Figure 3.6 (*Colour Section*) shows two prints comparing digital print and optical print.

3.3.2.3 Photofinishing

3.3.2.3.1 The photofinishing industry. During the past 35 years, photofinishing has grown into an industry of its own and has developed industrial equipment to do all the individual jobs, from film splicing and negative development to labelling and packing of the print orders. Besides large industrial units, the minilab has become established on the market with its share growing steadily during the last ten years (Table 3.2). Of course, this gives no indication of actual paper consumption, but the success of the minilab concept is self-evident. Economically speaking, the success of a photofinishing lab depends primarily on the quality of its logistic systems, its operators and efficient process control.

3.3.2.3.2 Printing. Industrial photofinishing is dominated by high-capacity printers capable of delivering 20 000 prints per hour. Film developing and printing (d&p orders) are integrated.

There are considerable differences in the design of processes, organization and equipment between wholesale labs and point-of-sale processors. Normally, low-quality film processing (e.g. due to over-cycled bleaches) remains unnoticed by the customer. The recognizable quality differences are surprisingly small.

The multiscanning printer (MSP) introduced by Agfa in 1985 enabled the printing yield to be raised to 99%. Total film scanning, the most efficient instrument for control of density and colour balance, has now

Table 3.2 The worldwide photofinishing market (1992) (Source: *Photofinishing News 1993*)

Source (units)	USA	Europe	Japan	Others	Total
National total	18 900	16 990	17 980	33 000	86 870
Minilabs	15 500	15 090	16 625	23 800	71 015*

*Represents 82% of the total.

been incorporated in innovative printer systems: the film is scanned at 140 measuring points (3 × 140 measurements per picture) before printing.

In a printer of the current generation, there is one white-light exposure only. In tungsten lamp bulb exposure, the colour balance is achieved by introducing two subtractive shut-off filters into the optical path during exposure.

In the minilab field, fully integrated systems comprising a film processing unit, a printing unit and a paper processing unit have become established. Printer and paper processor are combined in the form of a printer–processor.

Agfa's CRT printer, as the most advanced type of printer and the core of the Digital Print System introduced in 1990, provides access to high-quality prints with optimum colour saturation from colour slides.

3.3.2.4 Silver dye bleach materials (Schellenberg and Schlunke, 1976). Silver dye bleach materials are direct positive printing materials. They contain incorporated non-diffusing disazo or trisazo dyes of comparatively brilliant hue and very high light stability. When added to the emulsion layers, the dyes act as filters and reduce photographic speed. In a double-layer arrangement, where the dye layers are situated behind the emulsion layers, silver dye bleach materials are capable of exhibiting higher speed.

The only printing material on the market based on the silver dye bleach technology is Ilfochrome, formerly Cibachrome. Ilfochrome printing materials contain silver bromide-iodide high-speed negative emulsions (blue-sensitive, green-sensitized and red-sensitized) and three layers dyed in yellow, magenta and cyan, as well as one colour-correcting layer. Ilfochrome prints are noted for their outstanding lightfastness (index 6 on the ISO blue wool scale ranging from 1 to 8) and resistance to dark-fading. Wilhelm (1993) rates the shelf-life of a Cibachrome print as roughly five times that of a chromogenic print.

3.3.2.5 Amplification systems in colour printing. Two amplification systems for colour development have been proposed:

● the cobalt(III) complex amplification (Eastman Kodak); and
● the hydrogen peroxide amplification (Agfa Gevaert).

Although some technical problems of the hydrogen peroxide (H_2O_2) amplification process lie in the fugitive character of the peroxide-containing developer/amplifier solution and in the high vapour pressure of H_2O_2, amplification has been developed to a level at which it can be used on a practical scale. The material of choice for the hydrogen peroxide amplification system is a colour negative paper with silver chloride emulsions and low silver content. The amount of silver chloride needed is 10–20% that of normal colour negative paper if the resulting picture is to be free from granularity problems. At low silver coverage, the bleach fixing step can be replaced by a fixing step. Developed silver can be used to read out additional information with infrared scanning devices.

3.3.2.6 Instant photographic films and instant printing media. Instant photographic media require the formation of an image by the diffusion of image-providing substances from one layer to another. For a better understanding, an instant photographic material can be regarded as an integrated system of a negative film and a positive print. The donor layer and the receiving layer can remain in permanent contact, but they also can be kept in temporary contact only during development. The acutance of a diffusion print depends on the diffusion path between the image source (donor layer) and the image receiving layer. Integrated systems of film and print do not necessarily give better results than materials with a separate donor sheet. This is true of both black-and-white media and of instant colour photographic media.

3.3.2.6.1   Black-and-white silver salt diffusion systems (Schultze, 1979). Silver salt diffusion systems, as first developed by Gevaert in Belgium and Agfa in Germany, were based on diffusion reversal, a simple means for the one-step production of positives. Diffusion reversal involves the development of a negative by a developer containing definite amounts of a fixing agent in contact with an image receiving sheet (or layer) with incorporated metal or sulphide nuclei. While a negative image is built up in the donor layer residual, silver halide is dissolved and starts to deposit a positive silver image by physical development in the receiving layer. The physically developed silver in the positive exhibits high opacity, but its fading resistance is comparatively low. Today the silver salt diffusion principle is used in continuous tone and also line materials (for example, for phototypesetting).

3.3.2.6.2   Instant photography in colour: Dye diffusion systems. The basic idea behind dye diffusion systems is the transformation of an imagewise distribution either of the oxidant resulting from silver halide development or of residual developer (or reducing agent) into an imagewise distribution of diffusing dye. The variety of inventions within the framework of this concept is summarized by Van de Sande (1983).

'Dye developers' the first system of practical value, have been used in Polaroid instant colour materials since 1963. Dye developers consist of a dye residue and at least one hydroquinone developer residue which provides solubility in caustic alkali. They are diffusible only in their reduced state. Since silver halide development leads to the formation of non-diffusible oxidized dye developer, the dye is transferred only in low exposure areas. A dye developer in combination with a negative-type

Figure 3.4 (a) Spectral sensitivity of colour reversal film (printed on colour reversal paper). Some desaturation is recognised in the blue-green region. (b) Spectral sensitivity of colour negative film (XRG 100 Agfa). Negatives are printed on colour reversal paper (three different exposure levels).

(a)

(b)

Figure 3.6 Comparison of (a) conventional print from slide on colour reversal paper; with (b) Digiprint (print from slide by CRT-printer).

emulsion gives a positive dye image. The Polacolor system is discussed in detail by Bloom *et al.* (1978); a magenta dye developer is shown in section 3.4.3.2. The layer arrangement in Polacolor SX-70 is shown in Figure 3.7.

In its 'Spectra' material (Lambert, 1989), Polaroid uses a combination of dye developers for magenta and cyan and a yellow dye-releasing compound based on an argentolytic splitting reaction: the dye residue is split off from a ballasted thiazolidine carrier residue by reaction with silver ions from dissolved undeveloped emulsion grains. The latest version of the Polaroid instant colour imaging system (camera plus extended-range film) is called 'Vision'.

As a different approach, free of interlinking reactions, redox dye-releasing compounds with oxidative mobilization of image dyes have been developed by Eastman Kodak (Hanson, 1976) and Fuji Photo Film for use in instant colour films. Oxidative mobilization of image dyes leads to negative dye transfer. The reversal must be forced by the use of high-speed direct positive emulsions.

Former Eastman Kodak instant colour materials contained 4-sulphon-amido-1-naphthol derivatives as dye-releasing compounds, while Fuji's instant colour film Fotorama, introduced in 1981, contains ballasted 2-sulphonamidophenol derivatives. The chromophores are anions derived from simple monoazo dyes of no other use in dye chemistry. In particular, the magenta and cyan chromophores show the character of pH-indicator dyes.

IHR Dyes – based on a novel dye-releasing system with reductive dye-releasing characteristics – have been developed by Agfa Gevaert (see section 3.4.5.3) for the instant colour proof material Copycolor and for a single-sheet printing material, Afgachrome Speed (Peters, 1985).

Negative-acting sulphonamidophenol dye-releasing compounds and photothermographic silver salt dispersions sensitized by small amounts of silver halide constitute the dye-providing system in a semi-dry dye-diffusion printing material introduced by Fuji Photo Film within the 'Pictrography' system. All processing chemicals are incorporated into the imaging unit and external supply is restricted to heat and moistening water. The imaging unit comprises the light-sensitive donor film and the image-receiving sheet. It is exposed in the printer by passing through a trichromatic diode array (green, red, infrared).

A second, positive-working system, Fujix Pictrostat 100, was introduced in 1991. Further versions followed. The system as a whole, now called 'Pictrocolor', has been designed to provide quick access to copies or prints by a substantially dry, single-action process. Descriptions of the system are given by Shibata *et al.* (1988) and Iwano *et al.* (1994).

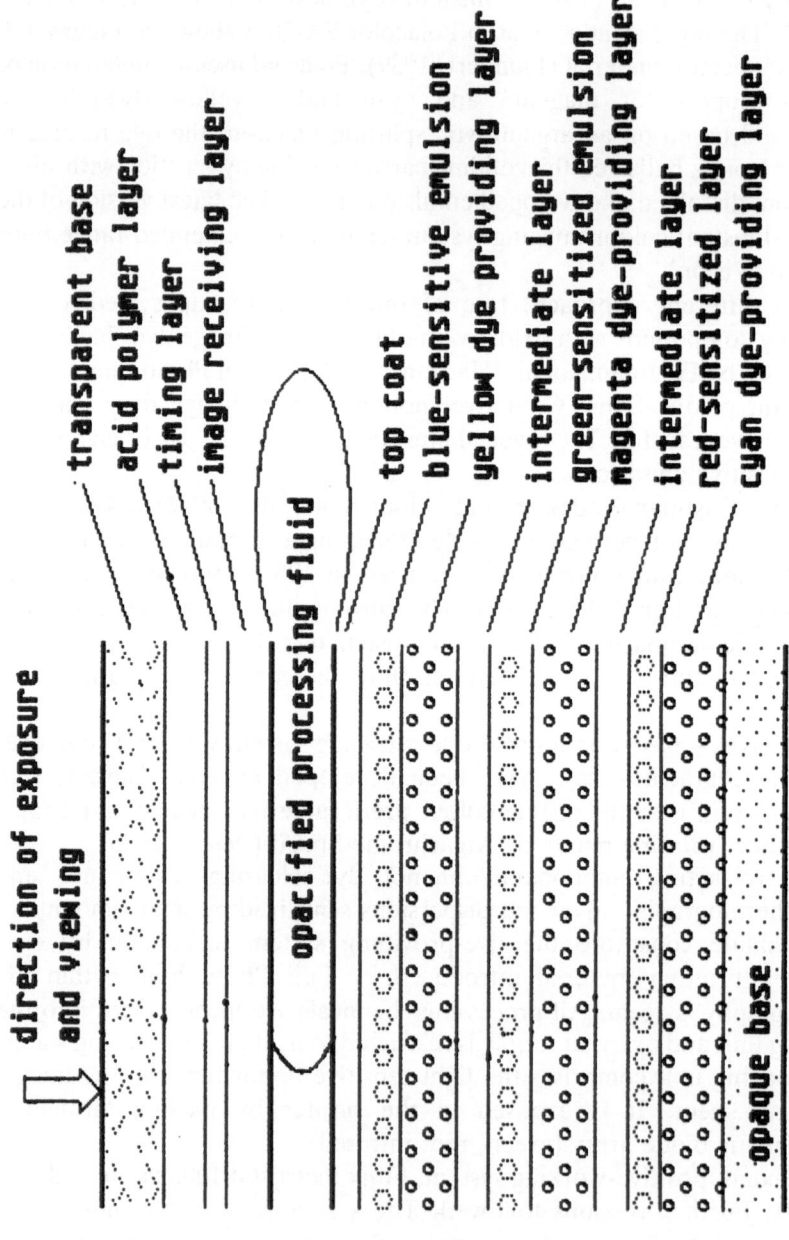

Figure 3.7 Polacolor SX-70. Layer arrangement.

direction of exposure and viewing

transparent base
acid polymer layer
timing layer
image receiving layer

opacified processing fluid

top coat
blue-sensitive emulsion
yellow dye providing layer

intermediate layer
green-sensitized emulsion
magenta dye-providing layer

intermediate layer
red-sensitized layer
cyan dye-providing layer

opaque base

3.4   The chemical constituents of photographic media

3.4.1   Emulsion ingredients and additives

A short review on sensitizing dyes has been prepared by Kampfer (1992); another article in *Research Disclosure* 17643 also contains some data on antifogging agents and stabilizers.

3.4.2   Developers and processing chemistry

3.4.2.1   Black-and-white development.   Black-and-white development has been well documented by Lee and Brown (1977). Neither hydroquinone nor the *p*-aminophenols have been displaced. Phenidone and its derivatives are still the superadditive electron-transfer agents of choice, and the value of reductones such as ascorbic acid as ecologically favourable developers has recently been recognized.

3.4.2.2   Dye formation: colour developers and chromogenic development.
In a colour photograph, the dyes are generated by a chemical process following reduction of exposed silver halide. Commonly used terms are 'chromogenic development' (Fischer, 1914) and 'colour development'. In chromogenic development, silver halide is developed by a colour developer of 4-amino-N,N-dialkylaniline type in the presence of a colour coupler. A treatise on current colour developers is given by Bent *et al.* (1951).

Chromogenic development is not the only kind of process capable of generating colour in a photographic layer from non-coloured compounds, but it is the only process of technical importance. The technical standards were defined by Eastman Kodak.

3.4.2.2.1   Colour film processing.   The processes for colour negative film (C-41) and colour reversal film (E-6 and successors) differ considerably in their chemistry as well as in their processing steps. In both processes, the colour developer still contains hydroxylamine and benzyl alcohol. The black-and-white developer of the E-6 process has to provide low D_{min} after colour development (Table 3.3).

3.4.2.2.2   Colour paper processing.   The RA-4 (Rapid Access) process for colour paper based on silver chloride emulsions clearly dominates the market. Strictly speaking, RA-4 stands for a group of processes which have in common that they all use a benzyl alcohol-free colour developer containing diethyl hydroxylamine as an antioxidant. The other components and processing steps are subject to variation (e.g. the use of separate

Table 3.3 CN and CR film development

Process	C-41			E-7*		
Colour developer	195 s	(38°C)		B&W developer	360 s	(38°C)
Bleaching bath	260 s	(38°C)		Washing	120 s	(33–39°C)
Washing	65 s	(38°C)		Reversal bath**	120 s	(33–39°C)
Fixing bath	260 s	(38°C)		Colour developer	360 s	(38°C)
Washing	195 s	(38°C)		Conditioner	120 s	(33–39°C)
Final bath	65 s			Bleaching bath	360 s	(33–39°C)
				Fixing bath	240 s	(33–39°C)
				Washing	240 s	(33–39°C)
				Stabilizing bath	60 s	

*Disclosed 1977.
**Contains Sn(II) salt as fogging agent.

Table 3.4 Rapid Access (RA) development

Process	RA-5	(RA-4)	R-3		
Colour developer	45 s	(35°C)	First developer	75 s	(38°C)
Bleach-fixing bath	45 s	(35°C)	Washing	90 s	
Washing	4 × 22.5 s		Second exposure		
			Colour developer	135 s	(38°C)
	Total: 3 min		Washing	45 s	
			Bleach-fixing bath	120 s	(38°C)
			Washing		

bleach and fixing steps or a single bleachfixing step; also washing steps which may be replaced by stabilizer treatment).

R-3 is an older type special process for colour reversal paper on the basis of silver bromide emulsions (Table 3.4).

3.4.2.3 The chemistry of colour development. At present, only toluene-diamine derivatives are in use as colour developers: CD-4(**1**) is more active while the more hydrophilic and less toxic sulphonamide derivative CD-3(**2**) gives more brilliant dyes.

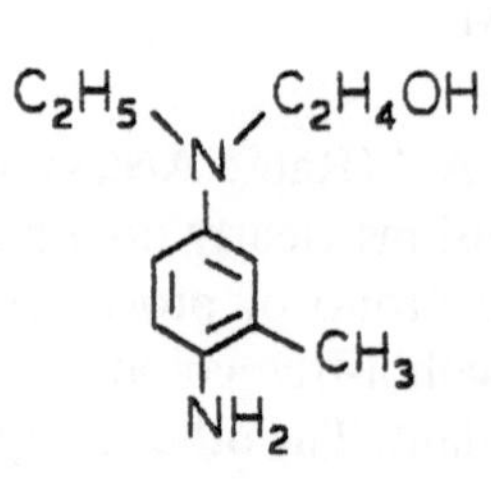

(1) 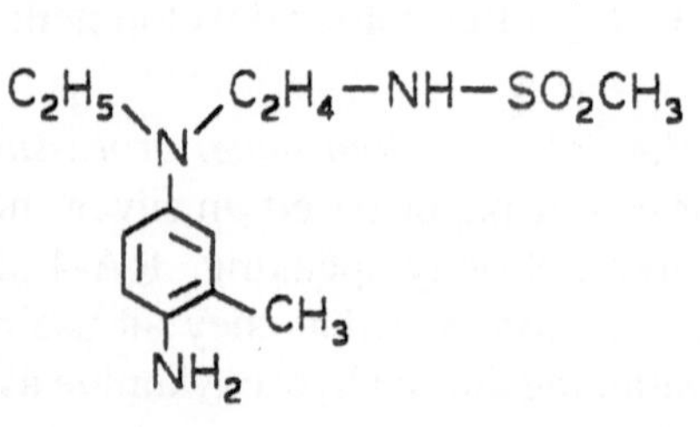   (2)

In the absence of a reactive coupler, colour developers are poor developers: their oxidation products are absorbed on developed silver and inhibit the passage of electrons. Oxidized colour developer is also believed to 'bleach' small silver nuclei on adjacent grains. These defects are not observed when the reactive type reacts with the carbanionic centre of a colour coupler.

The most reactive type in chromogenic development, the quinone diimine, is generated either by two-electron abstraction from the developer or by reversible disproportionation of semiquinone diimine, the radical primary product of development. In the absence of a coupler, the quinone diimine rapidly undergoes de-amination to a non-coupling quinone monoimine.

Coupler anions are the most powerful nucleophilic type present in the system and usually react first to form an intermediate leuco compound. Quinone diimines, on the other hand, are electrophilic reagents with high nitrene reactivity and also are dehydrogenating agents. The first step in colour coupling is usually regarded as an electrophilic substitution (Tong, 1977). Participation of radical reactions resulting from single electron transfer in the first steps of the coupling reaction is still being studied (Bent *et al.*, 1951).

Most of the colour couplers presently in use are of the two-equivalent type, which means that they contain a heteroatom group at the coupling position. In this case, the leuco compound formed in the first coupling step is transformed into the azomethine chromophore without participation of a dehydrogenation step, and dye formation may be described by the two simplified equations:

$$2AgBr + Developer + 2OH^- \rightarrow 2Ag + Quinone\ diimine \qquad (3.1)$$
$$Quinone\ diimine + Coupler \rightarrow Dye \qquad (3.2)$$

In the case of coupling reactions where hydrogen or an electrophilic substituent has to be split off from the coupling position (e.g. a $-COOH$ group), another molecule of quinone diimine is needed to dehydrogenate the leuco dye and the silver requirement rises to four equivalents per mole of azomethine dye.

Since two-equivalent couplers have acquired increasing technical importance and are likely to be used almost exclusively in the near future, only two-equivalent coupling is considered in detail. Scheme 3.1 shows the mechanism of two-equivalent coupling.

3.4.3 Colour couplers and special-purpose couplers (Tong, 1977; Wolff, 1992)

The couplers used in current colour films and in colour papers are hydrophobic substances from which – together with high boiling coupler

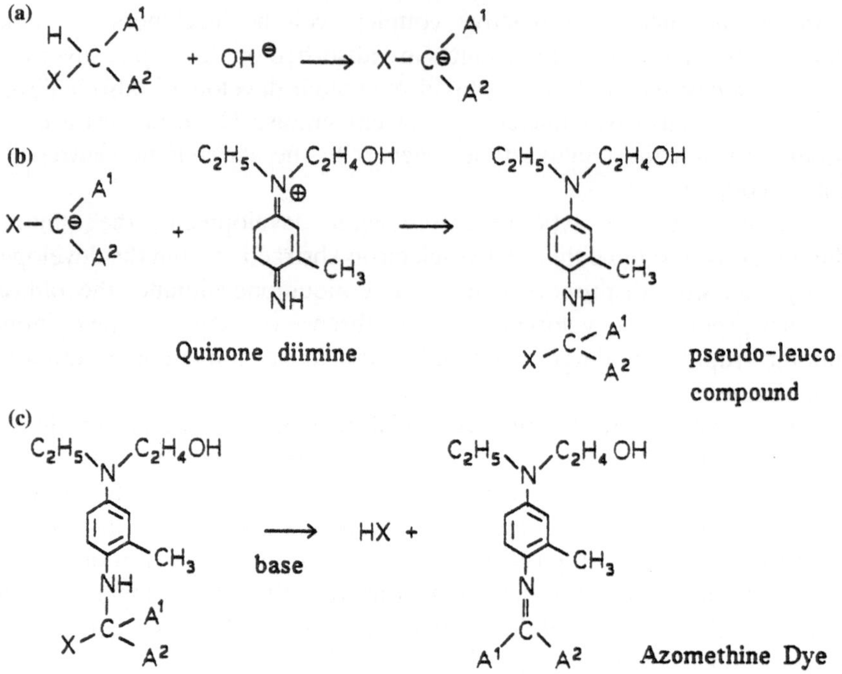

Scheme 3.1 Mechanism of two-equivalent-coupling: 3 Steps. (a) Deprotonation of coupler. (b) Coupling step. (c) Elimination of the leaving group. X = nucleophilic leaving group; A = electron accepting group.

solvents and optionally also stabilizing additives – oil-in-water dispersions are prepared and incorporated in the appropriate photographic layer. Latex couplers are used occasionally.

3.4.3.1 Yellow couplers. Until recently, only couplers of the benzoyl-acetanilide and pivaloylacetanilide classes were used as yellow couplers. Pivaloylacetanilides (3) produce azomethine dyes with their main absorption at a shorter wavelength than dyes derived from the benzoylacetanilide structure. To keep the hue of benzoylacetanilide azomethine dyes within the accepted range, the benzoyl group is substituted by a *p*-alkoxy substituent (4). The colour yield of benzoylacetanilides is higher, but they have a latent tendency to split at the benzoyl fragment under unfavourable conditions with loss of coupling activity.

3.4.3.2 Magenta couplers. In the past, different classes of compounds have been used as magenta couplers, e.g. indazolones (5) in colour paper and pyrazolo [5,1-a] benzimidazoles (6) in colour film. The most common magenta couplers are 3-acylaminopyrazolones (7) for colour film and 3-arylaminopyrazolones (8) mainly for colour paper. Almost all pyrazolone

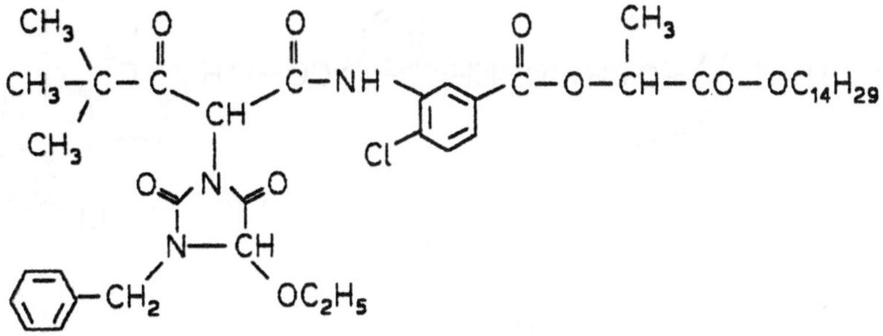

(3)

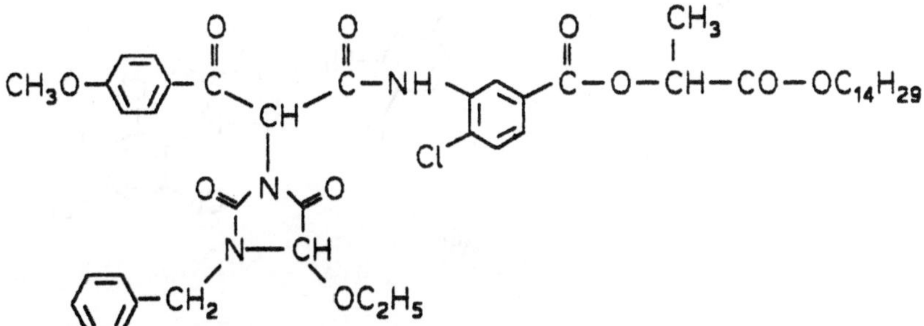

(4)

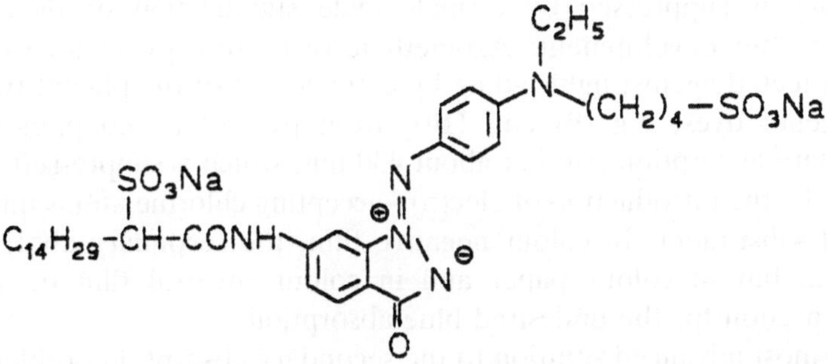

Dye from (5)

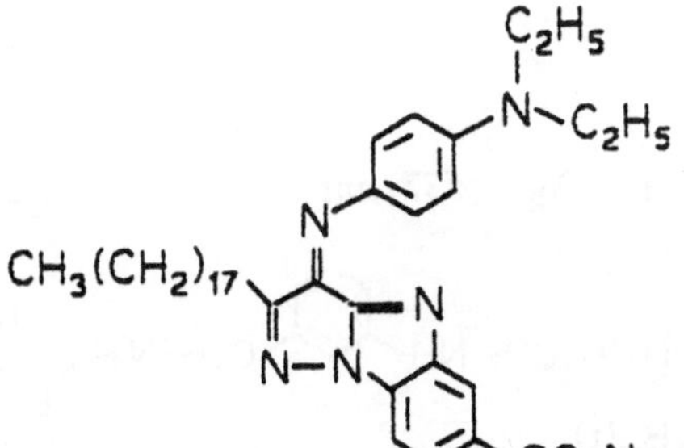

Dye from (6)

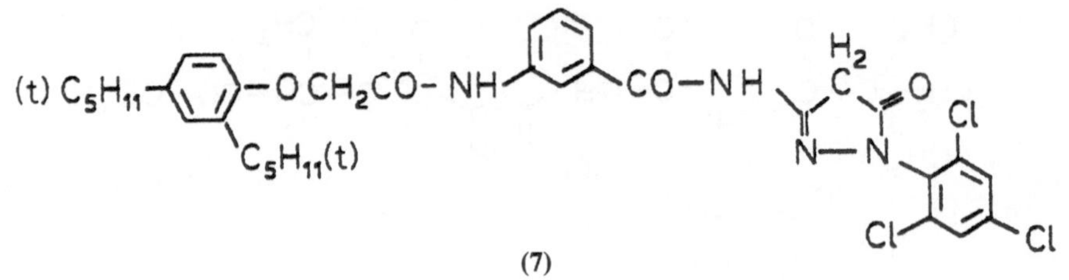

(7)

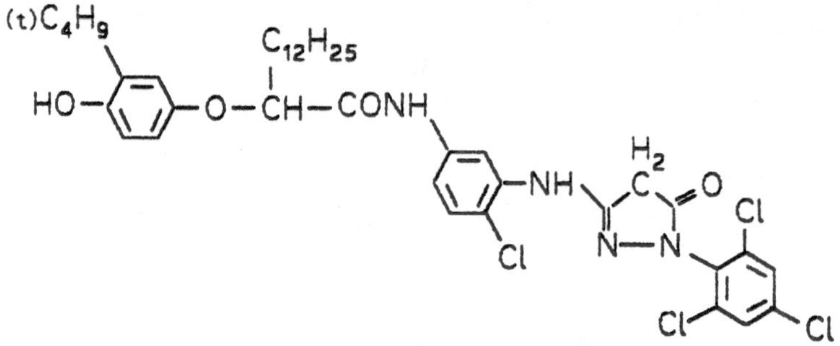

(8)

magenta couplers are notorious for their yellowing tendency under aging conditions and their sensitivity to aldehyde vapours. The yellowing tendency is suppressed by formaldehyde stabilization of the residual coupler after development. Azomethine dyes from pyrazolone couplers are protected against light fading by antioxidants of the phenol type.

Magenta dyes, e.g (9) and (10), from pyrazolone couplers show a secondary absorption band at about 430 nm, which is suppressed to some degree by the introduction of electron-accepting chlorine atoms into the 1-phenyl substituent. In colour negative film, the secondary absorption is masked, but in colour paper and in colour reversal film there is no compensation for the undesired blue absorption.

The most advanced solution to the secondary absorption problem came with the discovery of pyrazolotriazoles: pyrazolo [5,1-c][1,2,4]triazoles (11) and (12), and pyrazolo[1,5-b][1,2,4]triazoles (13).

Dye A: λ_2 *553 nm* λ_2 *437 nm*

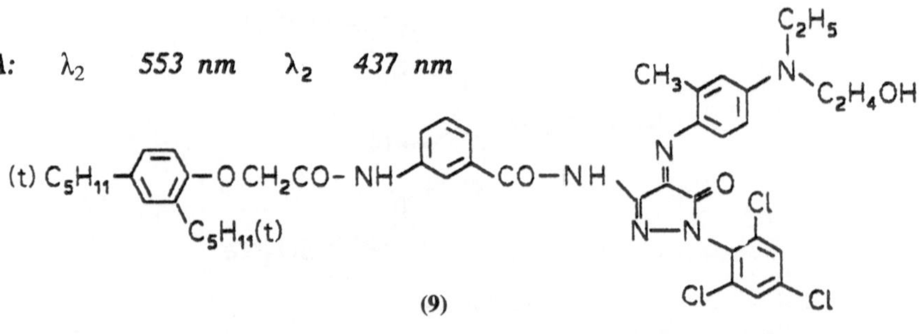

(9)

Dye B: λ_1: *544 nm* $\quad \lambda_2$: *430 nm*

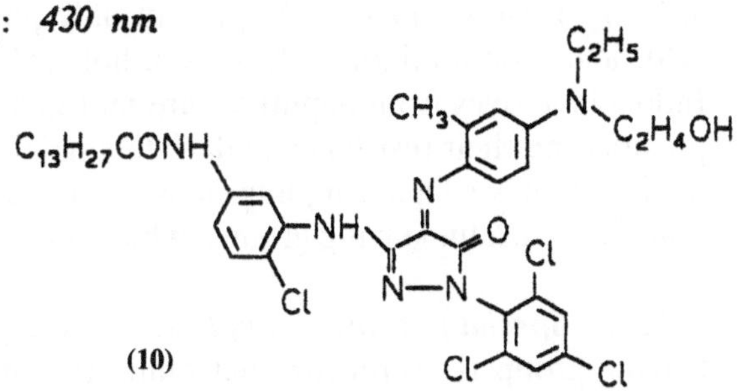

(10)

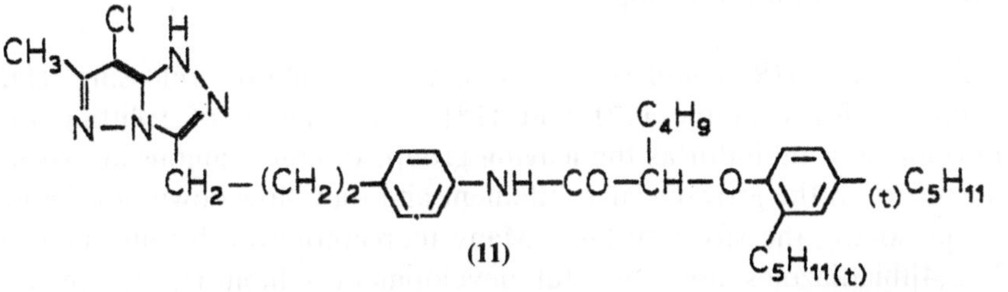

(11)

GB 1 252 418 (Bailey)

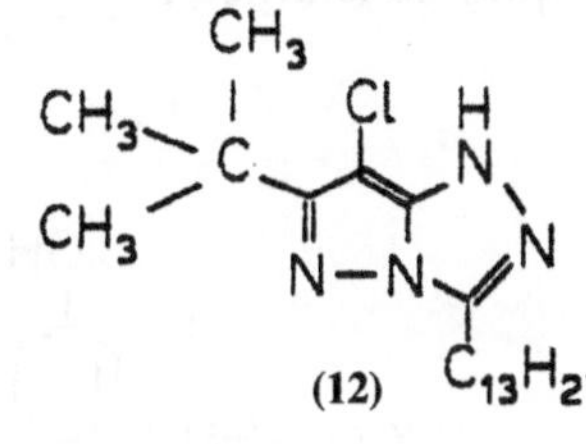

(12)

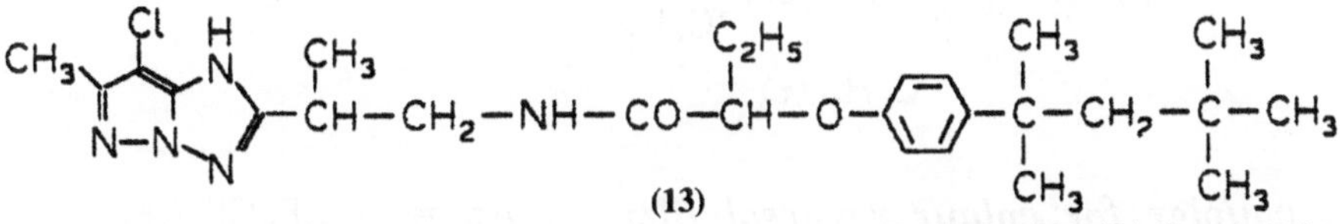

(13)

DE-OS 3 232 674 (Furutachi)

Both classes provide the brightest magenta azomethine dyes available for colour photography. As magenta couplers for colour paper, pyrazolo-triazoles give stable whites without any yellowing tendency. Serious drawbacks such as poor light stability were overcome by the introduction of a sterically demanding group at C6. This improvement is due to inhibited dye aggregation but is achieved at the cost of coupler reactivity.

The photooxidation of azomethine dyes from pyrazolo[1,5-b][1,2,4] triazoles and from pyrazolo[5,1-c][1,2,4]triazoles can be inhibited, for example by preventing free access of atmospheric oxygen to an exposed dye layer or by deactivation of the excited state. This is done by the addition of quenchers, for example electron-rich catechol diethers, to the coupler dispersion.

3.4.3.3 Cyan couplers. At present only phenols, for example (14) and
(15), and – on a minor scale – naphthols (16) are used as cyan couplers.
Indoaniline dyes from naphthols are more easily reduced than those from
phenols and their resistance to dark fading is comparably poor. In general,
naphthols have high coupling activity and are good carriers for photographically useful leaving groups, which may be split off from the coupler.

3.4.3.4 Special-purpose couplers. Two-equivalent couplers, whose
leaving group is useful for controlling the photographic process, can be
called special-purpose couplers.

  3.4.3.4.1 DIR couplers. Development inhibitor releasing (DIR)
couplers, for example (17) and (18), are capable of splitting off a
development inhibitor as the leaving group. Inhibitor anions are strongly
adsorbed on the growing silver filament and they slow down development
by poisoning the silver surface. Many mercaptotetrazoles and mercapto
[1,3,4]thiadiazoles are powerful development inhibitors, as are 1,2,3-
triazoles of moderate hydrophobic character.

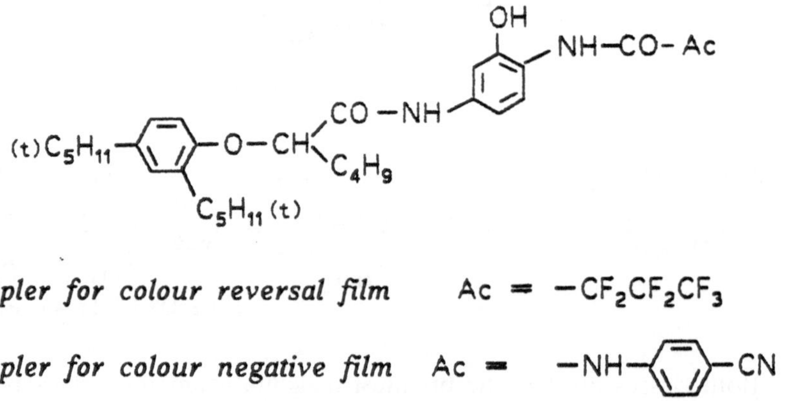

coupler for colour reversal film $Ac = -CF_2CF_2CF_3$

coupler for colour negative film   Ac = $-NH-\langle\rangle-CN$

(14)

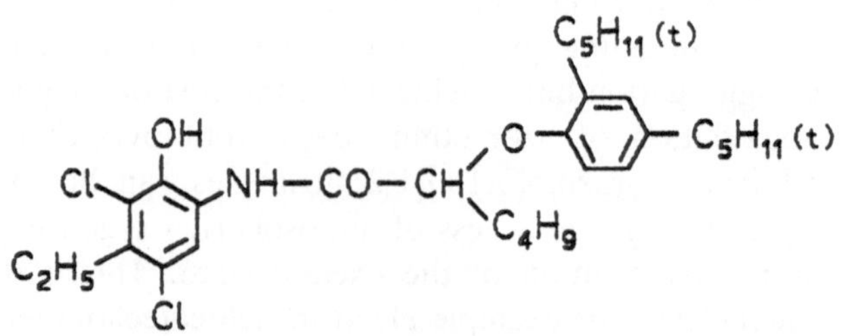

(15)

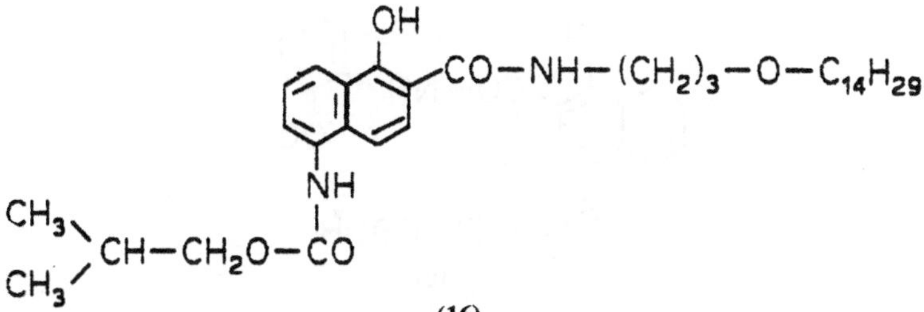

$$\text{OH}$$
$$\text{—CO—NH—(CH}_2)_3\text{—O—C}_{14}\text{H}_{29}$$
$$\text{NH}$$
$$\underset{\text{CH}_3}{\overset{\text{CH}_3}{>}}\text{CH—CH}_2\text{O—CO}$$

(16)

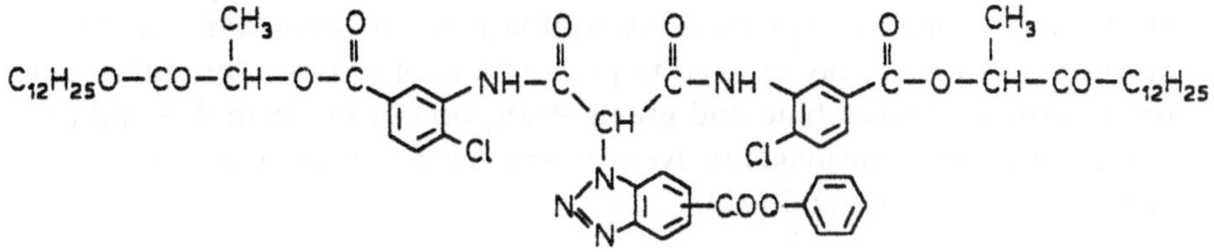

$$\text{C}_{12}\text{H}_{25}\text{O—CO—}\underset{\text{CH}_3}{\overset{}{\text{CH}}}\text{—O—}\overset{\text{O}}{\overset{\|}{\text{C}}}$$

$$\overset{\text{O}}{\overset{\|}{\text{NH—C}}}\text{—}\underset{\text{CH}}{\overset{}{}}\text{—}\overset{\text{O}}{\overset{\|}{\text{C}}}\text{—NH}$$

$$\overset{\text{O}}{\overset{\|}{\text{C}}}\text{—O—}\underset{\text{CH}_3}{\overset{}{\text{CH}}}\text{—CO—C}_{12}\text{H}_{25}$$

Cl   Cl

N
N—N

—COO—

(17)

$$\text{OH}$$
$$\text{—CO—NH—}$$
$$\text{OC}_{14}\text{H}_{29}$$
$$\text{S}$$
N
N—N
N

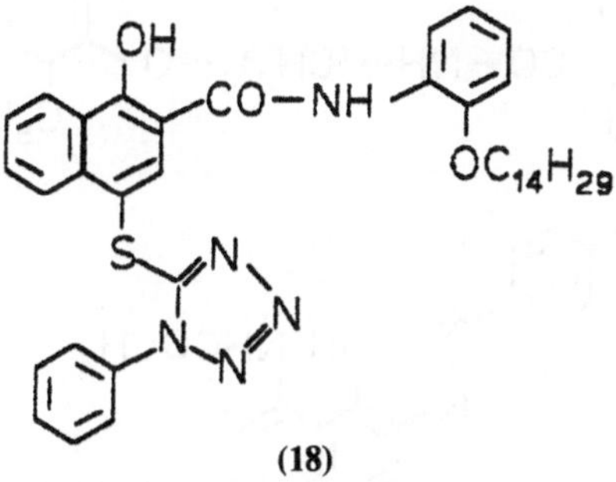

(18)

DIR couplers releasing development inhibitors of high diffusibility are responsible for chemical image processing by interimage effects and edge effects and they make the biggest contribution to high acutance, low granularity and enhanced colour contrast of modern colour negative films (Ranz, 1979).

3.4.3.4.2 DAR couplers. Development accelerator releasing (DAR) couplers and fogging agent releasing couplers (FAR couplers) are claimed to counterbalance the developer depletion and the inhibiting effect of bromide and iodide ions in the lower layers of a colour negative film and to improve the speed:granularity ratio.

3.4.3.4.3 BAR couplers. Bleach accelerator releasing (BAR) couplers are claimed to provide better bleaching after colour development, but they are also capable of activating development by forming dissolved silver complexes. A typical BAR coupler (**19**) splitting off β-mercapto-propionic acid is used in high-speed CN film.

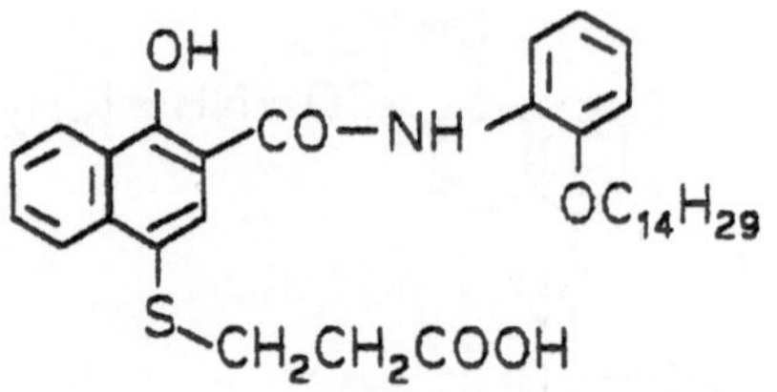

(19)

3.4.3.4.4 *Mask couplers.* Secondary absorption of image dyes is caused either by the slope of the main absorption band or by a separate electronic transition. Thus ideal absorption is rarely found and secondary absorption is favourably masked to prevent loss of colour information. At present the undesired blue and green absorption of the cyan dye and the blue absorption of the magenta dye are compensated by mask couplers – see structures (20), (21), and (22).

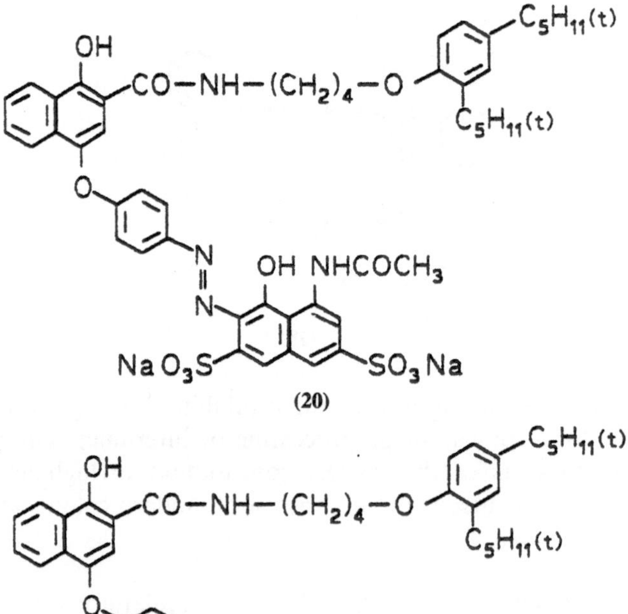

(20)

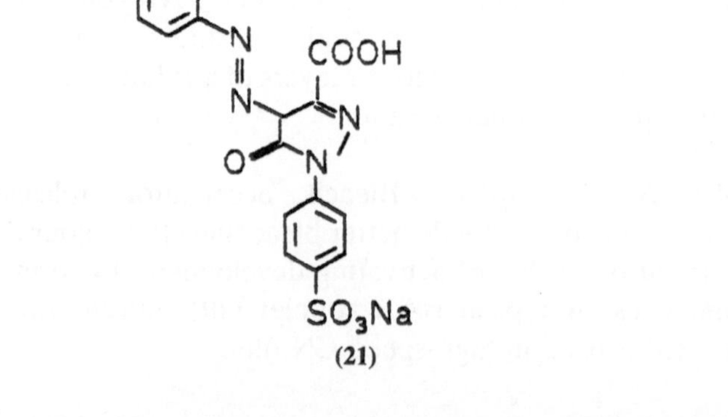

(21)

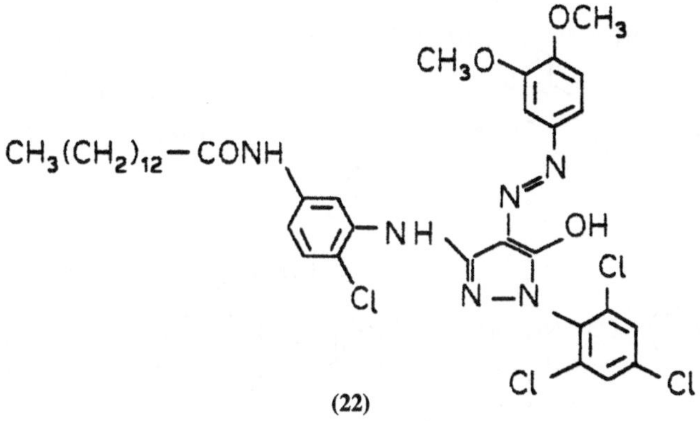

CH$_3$(CH$_2$)$_{12}$—CONH

OCH$_3$

CH$_3$O

N≋N

NH

OH

Cl

N—N

Cl

Cl — Cl

Cl

(22)

Since the masking of secondary absorption can also be achieved by balanced interimage effects, the use of mask couplers may decrease in the near future.

3.4.4 The chemistry of the silver–dye–bleach process

The concept of reductive dye bleaching (Schinzel, 1905) by developed silver formed the basis for some processes of technical importance, but only the Cibachrome/Ilfochrome azo dye bleach process is still in use (see section 3.3.2.1).

The silver–dye–bleach process makes use of the reducing power of developed silver towards azo dyes in sufficiently strong acids and in the presence of strong silver ligands (e.g. halogenide ions). Since the degree of dye bleaching is proportional to the amount of developed silver, the material provides positive prints from colour transparencies.

Ilfochrome's sulphonated azo dyes, e.g. (23) yellow, (24) magenta and (25) cyan, contain more than one azo group and are large non-diffusing molecules with a strong association and substantivity towards gelatin. In the dye–bleaching step, the chromophores are reduced to acid-soluble amino groups which render the colourless reduction products diffusible and enable them to be washed out. The established dye–bleach process comprises a developer, a combined dye bleach–silver bleach, a fixer bath and the necessary washing steps.

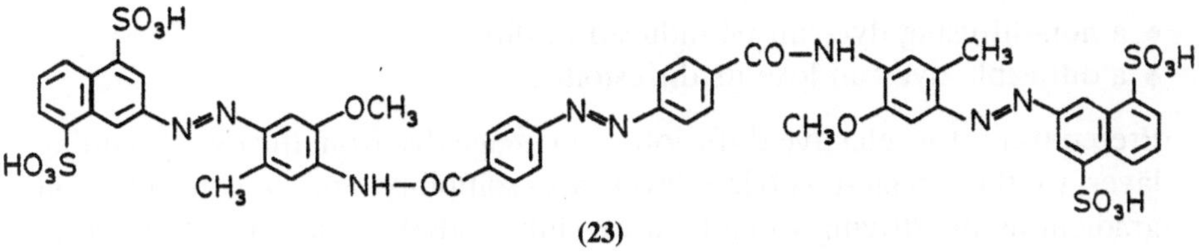

SO$_3$H

HO$_3$S

N≋N

OCH$_3$

CH$_3$

NH—OC

N≋N

CO—NH

CH$_3$

CH$_3$O

N≋N

SO$_3$H

SO$_3$H

(23)

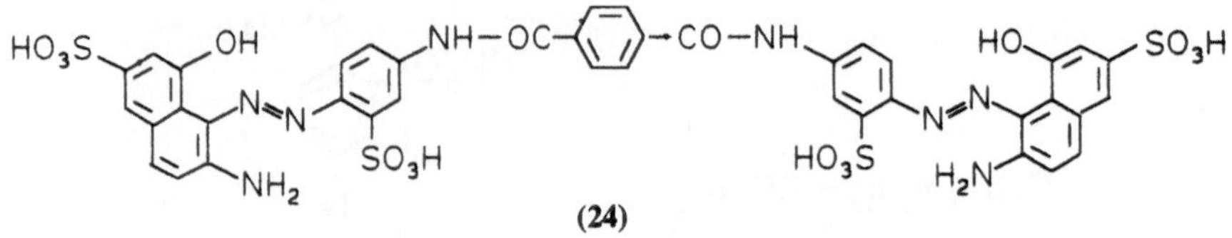

(24)

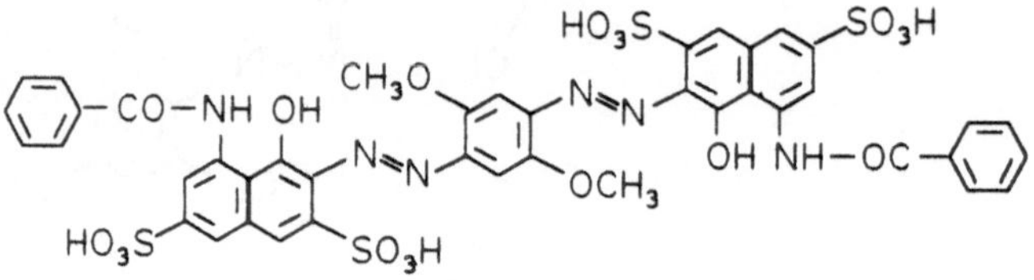

(25)

To enable the acid bleaching bath to function properly a bleach catalyst has to transport reduction equivalents (bound electrons) from the silver to the azo group of the non-diffusing dye. 1,4-Diazines, for example quinoxalines of well-controlled diffusibility, have been used as bleach catalysts to avoid deteriorated colour separation and acutance.

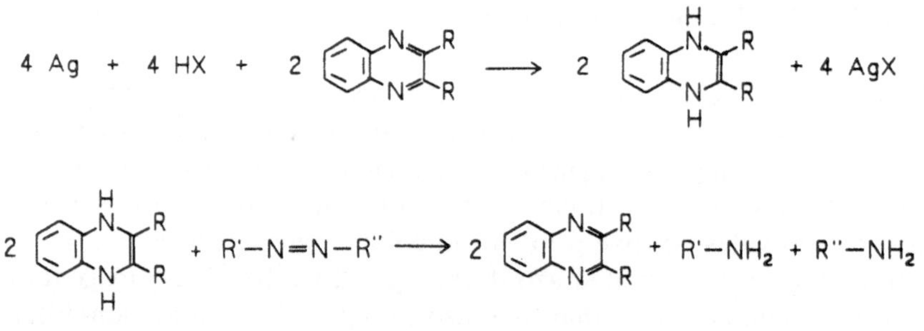

3.4.5 The chemistry of dye diffusion systems

Dye diffusion systems should be regarded as integrated systems of a film negative and a receiving element for the positive print. Basically, there are two possibilities of obtaining a dye image by the diffusion principle as the consequence of silver halide development:

- a non-diffusing dye can be induced to diffuse;
- a diffusible dye can lose its diffusibility.

Prerequisites for selective diffusion of an image dye from the dye-providing layer to the image-receiving layer are solubility and a concentration gradient as the driving force (source–sink model). Fixation of the dye in the image receiving layer is induced by the formation of a salt or an insoluble molecular complex between the dye and mordant.

If the image-receiving layer of an instant print remains firmly bound to the negative by the monosheet design of the material, provisions against post-process density increase, colour change and fringe formation have to be taken.

3.4.5.1 Dye developers (Polaroid). In a dye developer, the dye moiety is connected to at least one diffusion-controlling hydroquinone residue. Oxidation of the hydroquinone by electron transfer to the oxidized auxiliary developer renders the dye insoluble. An example of the dye developer chemistry is shown by the metallised magenta dye developer (**26**) used in the Polaroid SX-70.

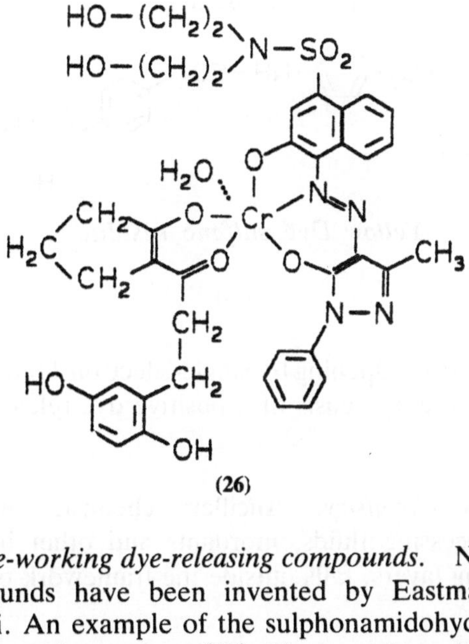

(**26**)

3.4.5.2 Negative-working dye-releasing compounds. Negative-type dye-releasing compounds have been invented by Eastman Kodak, Agfa Gevaert and Fuji. An example of the sulphonamidohydroquinonemono-ethers (Fuji) is shown in structure (**27**). The cyan dye sulphonamide is split off from the non-diffusing carrier residue by hydrolysis of a quinone monosulphonylimine after oxidation by an electron-transfer agent.

3.4.5.3 Positive-working dye-releasing compounds. As examples for positive-type dye-releasing compounds, IHR compounds (Agfa-Gevaert) and ROSET compounds (Fuji) deserve further mention. In an IHR dye (Increased Hydrolysis on Reduction), for example (**28**), the base-induced splitting of a dye–sulphone residue occurs with negligible speed in the quinonoid state of the carrier. It is activated when the carrier is reduced to the hydroquinone by an electron donor.

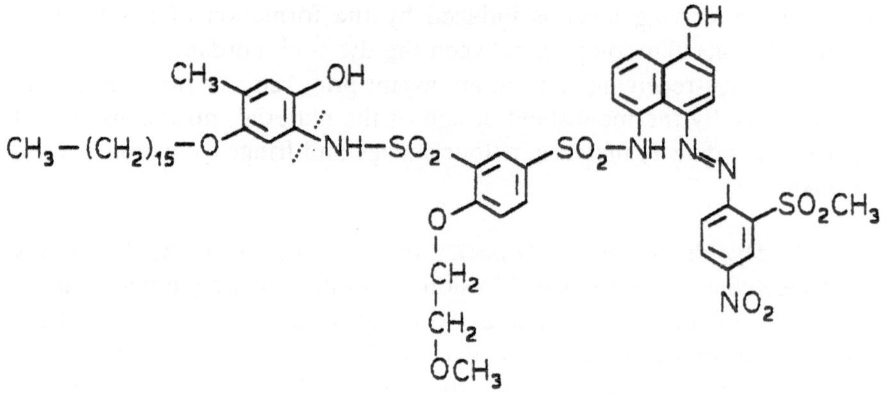

(27)   cyan dye-releasing compound

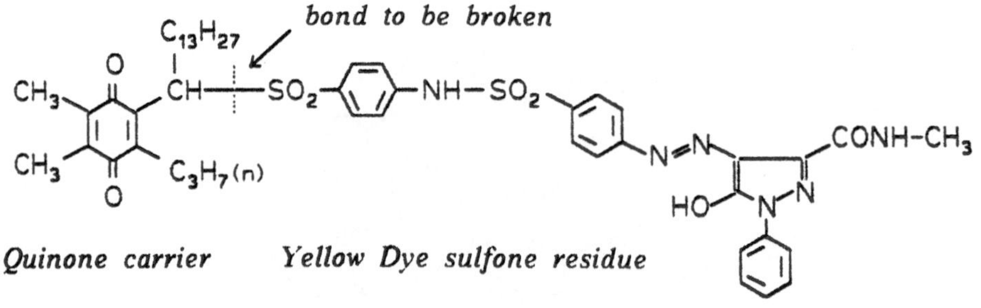

Quinone carrier   *Yellow Dye sulfone residue*

(28)

The principle of Ring Opening by Single-Electron Transfer (ROSET) was introduced by Fuji as the basis of a positive dye-releasing system (Fujix Pictrostat).

3.4.5.4 Ancillary chemistry. Ancillary chemistry, for example the chemistry of processing fluids, mordants and other ingredients of the process controlling layers, falls outside the framework of this chapter.

3.5 Photographic quality

Photographic quality has developed into the standard against which other imaging media are measured. The term is often regarded as a synonym for superb and unquestioned optical quality. Photographic quality may be equated with information content.

The major sources of information loss in photographic media are:

- light scattering (optical loss);
- density fluctuations (photographic loss); and
- diffusion phenomena (chemical loss).

Optical loss inevitably results from light scattering mainly on smaller grains. It is minimized by optimization of grain size and grain-size distribution. The same concept is valid for photographic losses, caused by inhomogeneous development within emulsion grain populations of different sizes and different dispersity of latent image.

Chemical information loss may have a very detrimental effect in colour development, where diffusion of oxidized colour developer – mainly of the more stable and less reactive semiquinone diimine – leads to diminished acutance and colour separation. In chromogenic development especially, the diffusion phenomena must be controlled carefully by optimizing coupler reactivity and process design. Within the material itself, the use of development inhibitor-releasing compounds and scavengers for the oxidized colour developer helps to minimize loss of information.

The theoretical treatment of definition and the discussion of practical methods for controlling and determining detail rendition are beyond the scope of this chapter but advances in chemical information enhancement as one of the sources of photographic quality should be mentioned: a deeper insight into the nature of the photographic edge effect (Nelson, 1971), a deeper understanding of interlayer effects, advanced emulsion technology and recent developments in coupler chemistry, have provided efficient tools for reconstructing lost information. Modern 35 mm colour films are specialized in resolving patterns of about 20 lines/mm which provide the highest contribution to the visual impression of acutance.

Apart from optical and photographic quality, it is the stability requirements which highlight on some well-recognized low points in photography:

- product stability problems; and
- process stability problems.

One fault, among many others, in colour photography is the insufficient storage stability of high-speed films, in particular under tropical conditions of high humidity or heat. In colour films exposed to high humidity, sensitizing dyes may begin to migrate from one layer to another, causing a breakdown of colour separation, emulsions may lose speed and become fogged and, apart from this, the characteristic curve for different colours may undergo non-linear changes, with the consequence of lost colour balance.

Quality problems in photographic prints are somewhat different from those in photographic film: in a print, definition is essentially determined by the master film. In a print of 9 × 13 cm from 35 mm film, a line pattern appears sharp from a distance of 30 cm when 20 lines/mm in the film are distinctly separated. One of the major causes of faults in colour paper printing is an inadequate chemical state of the process, for example induced by inadequate replenishment of colour developer or by over-recycling of the bleaches.

The problem area in colour paper, the core product of the market, is the stability of colour and whites. Wilhelm (1993) addresses 'the often troubled history of permanence in the color photography fields'. It must be emphasized, however, that the stability of colour prints from thermal dye transfer or from ink jet processes is at best equal to that of a chromogenic colour print. For chemical reasons, azomethine dyes exhibit a notorious acid lability and are sensitive to hydrolysis in general. Nevertheless, the light stability and dark-fading resistance of colour prints have reached a level where criticism by the consumer is no longer a major problem.

One new problem is insufficient bleaching in the production of photo-CDs, where residual silver reduces the scan rate by almost 50%, the slowest step in photo-CD production anyway. The ecological demand for closed-loop chemistry also creates new quality problems.

3.6 Future of photographic media

Many photographic media which played an important role in the past have now disappeared entirely; some have been replaced by better ones, others have vanished from the market without finding a successor. A good example of this is the chromogenic medical X-ray film; the collapse of the Super 8 amateur cine film system could also be mentioned in this context.

Common to all present considerations about the future of silver halide photography is the general belief that quality will determine its fate and that colour film will remain the superior imaging medium for a long time (Kriss, 1987). In the near future, the standard speed for CN film will be 400 ASA.

According to a frequently cited but not very meaningful figure, the photographic gain in a silver halide microcrystal amounts to 10^8. Ikenoue and Tabei (1990) calculate that the storage capacity of a 35-mm colour negative is about 60 times that of the CCD (charge-coupled device) chip of electronic still cameras. In addition, the CCD array has poor ability to cope with overexposure, and it can be demonstrated that increasing the storage capacity of a CCD chip to a value equivalent to HDTV (high-definition television) (the two-million pixel CCD) would result in a considerable loss of numeric aperture and a reduction of dynamic range. Besides this, the production yield of chips with increased storage capacity would decrease dramatically below the now accepted value of 10%. While these disadvantages will not change in the near future, there is still a high potential for increasing the performance of silver halide media, and colour film in particular.

3.6.1   Frontiers of the technically feasible

Digital printing has opened up new perspectives on the design options of an extremely simple amateur film for normal camera exposure.

The concept of the chromogenic colour film for scanning readout and printing is not new. Without elaborate image processing, a scan film cannot be used for the manufacture of colour prints, because its characteristic curve may be made extremely flat to extend the range of exposure and provide full security against improper exposure. Furthermore, the colour information may be read out separately from information on contrast or granularity. This can be achieved, for example, by using colour couplers of sufficiently different dye absorption to prevent the different information channels from mixing. It is impossible to predict, at present, to what extent the amount of silver halide in a scan film can be lowered to obtain more speed, better detail rendition and acutance.

Until recently only a few key patents covering the separation of information on colour, contrast and granularity existed, and the general aims is to eliminate coloured mask couplers and other light-absorbing media except the screening dyes.

Whether or not scan film will be able to occupy a segment of the film market, and consequently also of the photofinishing market, will depend on the economic success of fast scanning printing systems with either CRT exposure or with a laser array exposure system. Later modifications would depend on the use of differently sensitized colour paper.

The primary, indispensable prerequisite for this is a print output that can compete with the present generation of high-speed optical printers: a printing rate of 20 000 prints per hour is regarded as necessary. It must be doubted that the market will really accept a second standard of colour film alongside the established one if there were no additional advantages in either handling or economic reward.

3.6.2   Remote sensing and recording

Remote sensing and remote recording are creations of the space age. Everyone remembers the colourful pictures of the planet Jupiter taken from a scanning narrow angle camera on board Pioneer 11 and transmitted in digitized form. On close inspection this spectacular project reveals that the pictures were basically not photographic products in the classical sense.

Satellite photography, on the other hand, strongly depends on the use of photographic material which is sent back to Earth either by manpower or by means of containers recovered by aircraft once they have left orbit.

Worldwide, government administrations of all industrial and developed countries have discovered that remote imaging procedures are an excellent tool for controlling environmental changes in ecology, e.g. for control of seawater pollution along main naval traffic routes. The main problem of all photographic systems based on silver halide film and hardcopy is their inability to cope with the demand for real time data.

Wherever the highest possible information content is indispensable for localizing and defining environment changes, and when time-consuming photographic processing is not critical, airborne photography or photography from low-orbit vehicles may be as important as it is in military reconnaissance.

3.6.3   Digital storage of photographic images: a powerful tool for information management and archiving

Digital archiving is extremely helpful in reducing archiving space. The main problem of digital archiving is the need for updating the storage media periodically to keep the databases accessible over a long time by well-supported systems. The importance of long-lasting technical standards for digital storage has been accepted from the beginning, and the now established Photo-CD standard may represent the lower end of the market of a whole family of durable Write Once Read Many (WORM) times media.

For higher levels of digital storage, more specifically for digital storage or archiving of mass data, open standards are preferred and the optical disc technology seems to be firmly established. The technology of optical discs, a topic of military importance, is discussed in Chapter 7.

One critical point is rarely mentioned: while digital databases can be regarded as safe as long as they are filed on their storage medium, they are open to any kind of manipulation, which includes loss during copying procedures. Criteria for safe data management still have to be evolved.

It takes only a microscope to read data from a microfilm. From the point of view of silver halide photographic media, and especially the microform producers, it is regrettable that this important market may well be lost in the near future in spite of its clear independence from standards for electronic reading.

3.6.4   The merging of technologies: hybrid media and digital imaging

Medical imaging is an important tool for medical diagnosis and for documenting the state and results of medical treatment. It will continue to develop into a powerful market segment of imaging systems. The way technical development is going would seem to be clear. Most systems now in use – except X-ray photography – generate digital data first which are then transformed into a series of pictures on a screen for better evaluation. Prints can be made as required, but not often on photographic paper.

More and more physical investigation methods provide information that either describes or necessitates local distribution of quantified data that is worth documenting. Therefore, pictures will be preferred to listed data in handling surveyed information, e.g. in perspectivic radar, high-resolution thermography, high-speed kinematography and other technical applications.

The same is true of certain medical applications like normal medical ultrasound imaging or ultrasound colour-coded Doppler imaging, computer tomography or *in vivo* magnetic resonance imaging for observation of tissue function. Immediate availability of pictures as documents is only possible either with expensive instant photographs of the screen image or by printing. The time when high-quality coloured prints from full-colour screens were more expensive and time-consuming than photographic prints may already have passed – see Chapters 5 and 6.

Business graphics is another field which clearly illustrates the future perspectives of digital imaging. By now digital imaging is developing into a new field of business for – amongst others – photographic media.

Photographic media make up the upper end of the market where data from digitally generated or processed pictures are transmitted from disc to film by scanning exposure and conventional continuous tone colour prints are made in professional laboratories (Figure 3.8). Imaging studios are hesitant to take over the whole job, while PRO labs, on the other hand, apparently avoid adding activities with unsafe returns and high staff requirement. Magneto-optical discs could be the data-storage media of choice to store the large databases needed (see Chapter 7). A need for new photographic products is not a priority.

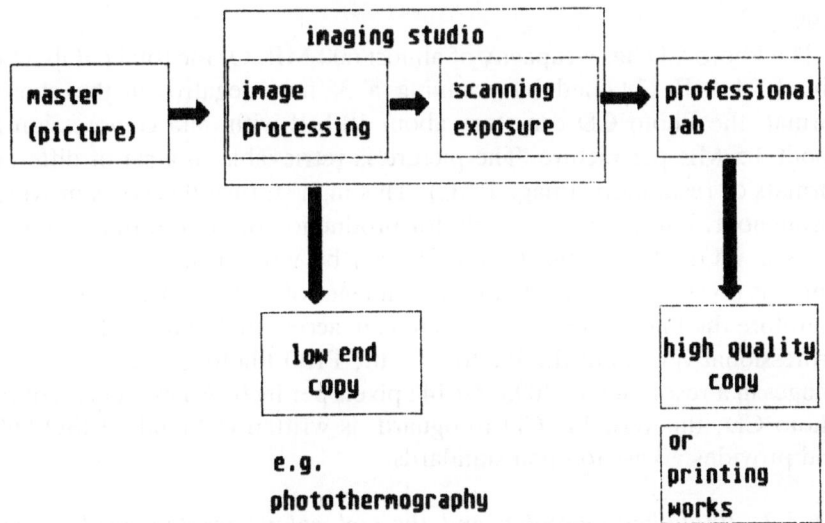

Figure 3.8 Hybrid images for specific markets.

3.6.5   The merging of uses (communication and entertainment)

Very little is known about the home-printing and imaging market. Experiments by Eastman Kodak and Agfa Gevaert around 1985 to open up a latent market with new dye diffusion materials like Ektaflex and Agfachrome Speed have not turned out very well. In the meantime, however, the interests of customers have changed and the home PC has encouraged younger people in particular to put their creative abilities to the test and to play with images at the lower end of the market. What the new procedures provide is an unexpected freedom of creativity.

The Photo CD and innovative systems for office printing or the home printing market (e.g. thermal dye transfer video prints or prints on either semidry colour photo thermographic material Fuji Pictrography or Pictrostat, or on the new Fuji Thermo Autochrome) have created fresh possibilities for this activity. Two features are essential within the market: firstly, the handling and safety standards for the equipment must be high, and secondly, the procedures do not meet critical demands in speed and have to be carried out without chemical spill.

3.6.5.1   The photo CD.

The Photo CD was created originally to store comparatively large numbers of colour slides and/or colour negatives in digital format without any recognizable loss of information: the pictures must be stored on easily accessible standarized storage media of good stability and be presentable for home use applications on TV screens. Of course, the reading procedure on a standard CD-ROM drive opens the digital database from a Photo CD to any kind of processing on a personal computer. The technical design of the system is described by Marchesi (1994).

The Photo CD has a capacity of almost 600 MB. Of the total database of about 24 MB obtained by scanning a 35 mm negative in the chosen format, the Photo CD can store about 18 MB with data compression to about 4.5 MB per picture. The picture is retrievable in a set of different formats of resolution (Image PAC). This high storage density is provided throughout, but it is needed only for production of large-format prints.

As a WORM disc, the Photo CD can be written in several sessions. There is no cheaper access to digitized images of high-quality available and therefore the Photo CD is expected to gain access to desktop publishing. A professional version of the Photo CD, the PRO Photo CD, can also store images in a resolution of 4096×6144 pixels per inch. A new version of the Photo CD, the Writable CD Infoguard, is written in Standard ISO 9660 and provides access to open standards.

3.6.5.2   Photothermographic and thermal colour printing media.

The presentation of the Fuji Thermo Autochrome material in February 1994 marked the realization of a novel concept. The older photothermographic instant printing systems of Fuji Pictrocolor use a light-sensitive material based on silver halide sensitized organic silver salts and dye-releasing chemistry (see section 3.4.5).

The new Autochrome system avoids even the inconvenience of light sensitivity and permits processing under roomlight conditions. A set of microencapsulated dye-providing compounds in three grades of thermal sensitivity is incorporated into the material, which consequently has to be activated in three successive thermal writing steps at three different temperature levels. After each writing and dye-mobilizing step, the non-activated areas must be shut off by ultraviolet curing of the residual microcapsules (cf. Cycolor process, Chapter 6). Of course, the multistep printing process is very slow and the delivery of one print takes 90 seconds. However, the quality is said to be very close to the quality of a photographic print and the system may be superior to other thermal dye transfer printing systems.

3.6.5.3 Future technologies capable of replacing the chromogenic colour print. From today's viewpoint, only two systems have the technical capabilities to push the colour paper prints from the market:

- an electrophotographic process to replace the smaller formats (see Chapter 4); and
- a high-resolution ink jet printing process to replace the large formats (see Chapter 5).

Until now, both technologies have been performing well in serial printing but this situation may change. The challenge in small-format printing lies in the need for extreme speed, for example 20 000 prints per hour as in industrial finishing, at low cost and high quality throughout. Until now, there have, however, been serious limitations to light stability. Half-tone quality problems seem to have been solved.

3.6.6 Ecological challenges of photographic processing

The photographic industry will have to meet some important challenges in the near future.

- The degree of silver recovery will have to be raised further, from about 90% to possibly 99%. This can be done by two-step extraction from bleaches and wash water (metal reduction plus ion exchange).
- The discharge of ammonium ions will have to be reduced.
- The problem of bromide and iodide discharge/recovery must be tackled.
- Discharge of oxidizable organic substances must be minimized.

● Closed-loop/zero-discharge processes will be established. This means that water will be recovered, although at high energy cost (spill evaporation) and only solid waste will be generated.

3.6.7 The advanced photo system

Together with four important Japanese manufacturers, Eastman Kodak has announced the development of a new photographic system involving smaller cameras, 'smart films' of higher detail rendition, and new processing equipment. The Advanced Photo System is designed to coexist with the 135 system and eventually to take its place. The film will have a magnetic backing of low optical density. Its format will be slightly smaller than the current 24 × 36mm and will provide about 60% of its picture area.

The magnetic backing will record and exchange additional coded information useful for improving print quality and store personal data and messages from the camera user. It could be doubted whether an additional possibility of making partial prints will be compatible within the routine of high-speed industrial printing.

The basic concept of hybrid media must contribute to user comfort and satisfy the consumer's claim for quality. For this purpose, provisions for safe storage of the negative will also be made and repeat orders will be carried out without loss of quality.

The future of amateur photography will, however, not depend on the success of this project. The consumer's decision for either a hardcopy or the more volatile video media will be a question of culture, aesthetic standards and viewing habits.

References

Bent, R.L., Dessloch, J.C., Duennebier, F.C., Fassett, D.W., Glass, D.B., James, T.H., Julian, D.B., Ruby, W.R., Snell, J.M., Sterner, J.R., Thirtle, J.R., Vittum, P.W. and Weissberger, A. (1951) *Journal of the American Chemical Society*, **73**, 3100.

Berry, C.R. (1977) Precipitation and growth of silver halide emulsion grains. In *The Theory of the Photographic Process* (4th edn), T.H. James (ed.). Macmillan, New York.

Bloom, S., Green, M., Idelson, M. and Simon, M. (1978) The dye developer in the Polaroid photographic process. *Chemistry of Synthetic Dyes*, No. 8, 331–337.

Ducos du Hauron, L. (1869) A new heliochromic process. *The Photographic News*, **13**, 319.

Fischer, R. and Siegrist, H. (1914) *Photographische Korrespondenz*, **51**, 18–22.

Gilman P.B. (1977) *Pure and Applied Chemistry*, **49**, 35.

Hanson, W.T. Jr. (1950) *Journal of the Optical Society of America*, **40**, 166–171.

Hanson, W.T. Jr (1976) A fundamentally new imaging technology for instant photography. *Photographic Science and Engineering*, **20**, 155–160.

Heinen, W. (1988) Printen wie das Auge sieht. *Fotowirtschaft: Thema des Monats*, December 1988, 6–10.

Hunt, R.W.G. (1988) *The Reproduction of Colour* (4th edn). Fountain Press, Tolworth, England.

Ikenoue, S. and Tabei, M. (1990) Colour negative film technology compared with that of an electronic still camera. *Journal of Imaging Science*, **34**, 187–196.

James, T.H. (1986) Chemical sensitization, spectral sensitization and latent image formation in silver halide photography. *Advances in Photochemistry*, Vol. 13, 239–425. Wiley-Interscience, New York.

Krause, P. (1989) *Imaging Processes and Materials*, Neblettes's 8th edn. J. Sturge, V. Walworth and A. Shepp) (eds), pp. 110–118. Van Nostrand Reinhold, New York.

Kriss, M.A. (1987) *Journal of the Society of Photographic Science and Technology Japan*, 50, 357.

Kampfer, H. (1992) *Spectral Sensitization: Ullmann's Encyclopedia of Industrial Chemistry*, Vol. A20. *Photography*. VCH Verlagsgesellschaft, Weinheim.

Lambert, R. (1989) A technical description of Polaroid's Spectra film, *Journal of Imaging Technology*, 15, 108–113.

Land, E. (1977) *Photographic Science and Engineering*, 21, 225–236.

Lapp, O. (1992) *Emulsions: Ullmann's Encyclopedia of Industrial Chemistry*, Vol. A20. *Photography*. VCH Verlagsgesellschaft, Weinheim.

Lee, W.E. and Brown, E.R. (1977) The developing agents and their reactions. In *The Theory of the Photographic Process* (4th edn), T.H. James (ed.), pp. 291–334. Macmillan, New York.

Liggero, S.H., McCarthy, K.J. and Stella, J.A. (1984) *Journal of Imaging Technology*, 10, 1–9.

Malin, D.F. (1993) Science in pictures. A universe of colour. *Scientific American*, 269, August 1993, 56–61.

Marchesi, J.J. (1994) Professionelles Rund um die Photo CD. In *Photographie + plus +*, 1 April 1994, 104–109.

Matejec, R. (1992) Theory of development. In *Ullmann's Encyclopedia of Industrial Chemistry* Vol A20, *Photography*. VCH Verlagsgesellschaft, Weinheim.

Maxwell, J.C. (1858–62), *Proceedings of the Royal Institute*, 3, 370.

Mitchell, J.W. (1993) The silver halide photographic emulsion grain. *Journal of Imaging Science and Technology*, 37, 331–343.

Nelson, C.N. (1971) *Photographic Science and Engineering*, 15, 82.

Iwano, H., Ozaki, H., Yokokawa, T. and Inagaki, Y. (1994) New type photothermographic systems for rapid access colour prints. *IS&T's Eighth International Symposium on Photofinishing Technology, Final Program and Advanced Printing of Paper Summaries*. IS&T, Springfield, Virginia.

Parsons, T.F., Gray, G.G. and Crawford, I.H. (1979) To RC or not to RC? *Journal of Applied Photographic Engineering*, 5, 110–117.

Peters, M. (1985) A novel dye diffusion material for prints from colour slides. *Journal of Imaging Technology*, 11, 101–104.

Ranz, E. (1979) *Chemie für Labor und Betrieb*, 30, 229.

Rogers (1961) US Patent No. 2 983 606.

Schellenberg, M. and Schlunke, H.P. (1976) *Chemie in unserer Zeit*, 10, 131–138.

Schinzel, K. (1905) *British Journal of Photography*, 52, 608.

Schultze, D.F. (1979) Graphic arts applications of silver diffusion transfer systems. *Journal of Applied Photographic Engineering*, 5, 163–166.

Shibata, T., Sato, K and Aotsuka, Y. (1988) Full colour printing system based on photothermographic material (2). *Paper Summaries, 4th International Congress on Advances in Non-Impact Printing Technologies*, New Orleans, March 20–25, pp. 358–361 SPSE, Springfield, Virginia.

Siegel, J. von Grossmann, J., Besserdich, H., Heinrich, B., Stösser, R and Sydow, M. (1990) On the interaction between chemical and spectral sensitization. *Journal of Photographic Science*, 38, 47–53.

Tani, T., Kikuchi, S. and Honda, K. (1968) *Photographic Science and Engineering*, 12, 80–89.

Tong, L.K.J. (1977) Mechnism of dye formation and related reactions. In *The Theory of the Photographic Process* (4th edn), T.H. James (ed.). Macmillan, New York, New York.

Van de Sande, Ch. (1983) Farbstoffdiffusionssyteme in der Farbphotographie. *Angewandte Chemie*, 95, 165–256.

Vogel, H.W. (1873) *Berichte der Deutschen Chemischen Gesellschaft*, 6, 1302.

Wilhelm, H. (1993) *The Permanence and Care of Color Photographs*. Preservation Publishing, Grinell, Iowa.

Wolff, E. (1992) *Ullmann's Encyclopedia of Industrial Chemistry*, Vol. A20, *Photography*. VCH Verlagsgesellschaft, Weinheim.

4 Electrophotography

R.S. GAIRNS

4.1 Introduction

Electrophotography, as the name suggests, is a process in which an image can be reproduced by means of electricity and light. Literature detailing examples of electrophotographic processes date back to the early 1920s. Initial attempts to develop photo-electrical imaging processes were centred on the use of electrosensitive papers which could be activated by photocurrents (Von Bronk, 1922). Commercially, the most significant invention in the field of electrophotography was the discovery of the process most commonly known as xerography. Xerography, derived from the Greek, means 'dry writing' and was developed out of the original inventions by Chester Carlson (Carlson, 1938).

It had long been known that certain materials could be made to accept an electrostatic charge and that this charge could be used to attract fine particles. Carlson noted that certain materials, photoconductors, when exposed to bright light would lose their charge and he was able to demonstrate that these materials could form the basis of a device that could be used to produce an electrostatic image which, when developed with powder, could be transferred to paper. In Carlson's experiments, plates made from amorphous sulphur were constructed by casting molten sulphur on a conductive metal base. Carlson charged his plates by rubbing with a cloth; following exposure the plate was developed using a fine powder.

The development of these early experiments into a reliable process which could be utilised commercially was undertaken at the Battelle Memorial Institute, Columbus, Ohio. Research began in 1944 and through funding from the Haloid Company (later to become the Xerox Corporation) and the US Army Signal Corporation, the techniques were developed (Dessauer et al., 1955). In 1950 the first commercial xerographic equipment was launched.

4.2 The technology of xerography

Schaffert, in his book on electrophotography, attempted to define the technology in terms of the nature of the latent image formed (Schaffert,

1975). He also offers a detailed explanation of many of the techniques which have been published and patented. In this chapter, only those techniques which have been the most commercially successful are considered.

Xerography is arguably the most significant development in the field of electrophotography. Carlson utilised the photoconductive properties of materials such as sulphur and anthracene to establish the basic principles. These findings indicate the basic stages of the process (Figure 4.1):

1. The photoconductor is first charged in the dark.
2. The plate is then partially exposed to illumination in order to produce an electrostatic image.
3. The image is then developed with a fine powder.
4. The developed image is transferred to the receiving substrate (paper, cloth, overhead transparency, etc.)
5. The image is fixed to the substrate.
6. Discharge and cleaning of the photoconductor.

Although the basic concept proposed by Carlson had to undergo considerable refinement to produce a commercially viable product, modern photocopiers and laser printers still operate by the basic process established in the 1940s. Many of the major developments have been in the engineering and mechanics of the process and the consumables utilised by these devices.

4.2.1. Charging

Originally, Carlson made use of simple friction to obtain an electrostatic charge on his photoconductor. In modern copiers and laser printers, high-voltage devices are used to produce a very even distribution of charge on the photoconductor. The most common charge device is the corona which consists of a fine wire held close to the photoconductor (Dessauer et al., 1955). When a high voltage (6 kV) is applied to the wire and the photoconductor base is grounded, the wire causes ionisation of the air and the charged gaseous molecules are drawn to the photoconductor surface by the applied field. By passing the photoconductor under the corona at constant velocity, a uniform charge is applied to the surface.

A variation on the corona is the scoritron, which has a grid wire between the corona wire and the photoconductor. A smaller voltage is applied to the grid and this acts to provide a more even charging of the photoconductor and prevents overcharging of the surface. The presence of high voltage fields can lead to the production of ozone and oxides of nitrogen being formed in close proximity to the photoconductor which can result in surface damage. Furthermore, the emission of these gases cause undesirable odours and is of environmental concern. In some systems conductive

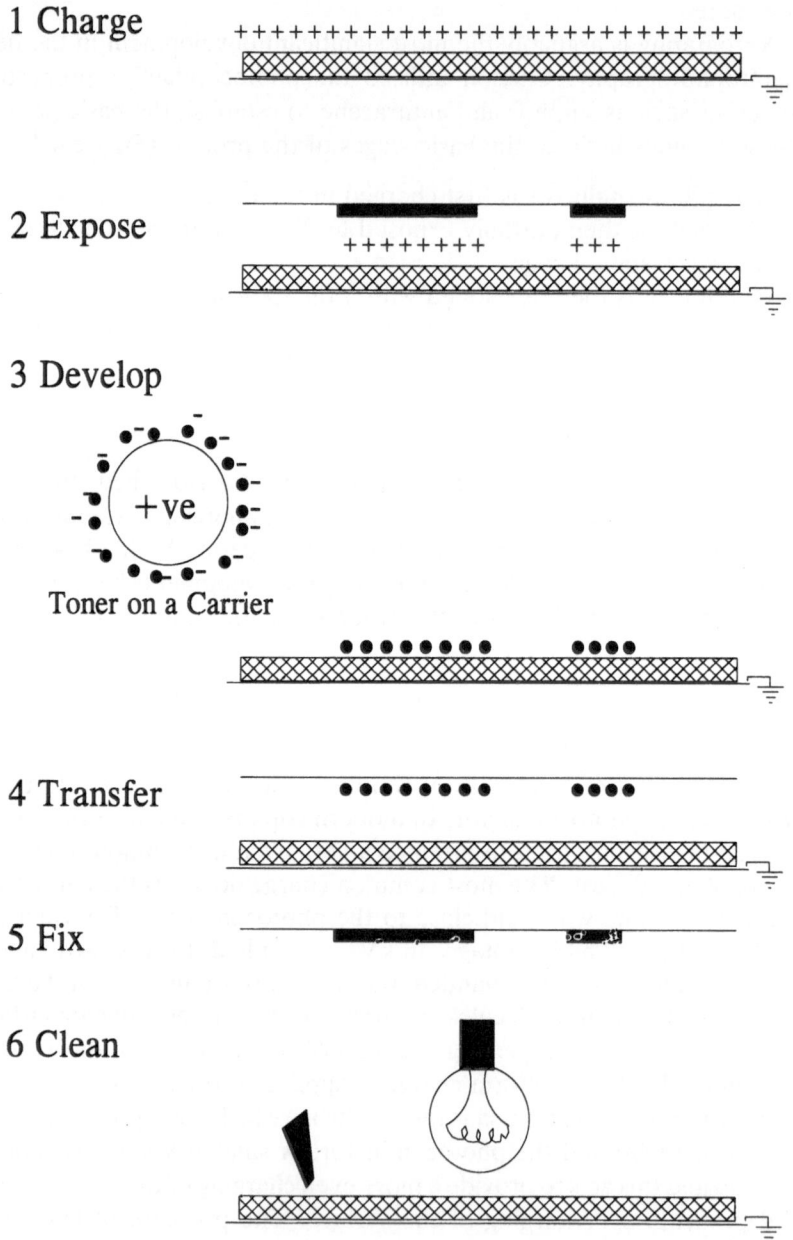

Figure 4.1 The electrophotographic process steps.

rubber rollers are used to apply a uniform charge (Straughan and Mayer, 1958). These work by direct contact with the photoconductor and carry much lower voltages.

4.2.2. Exposure

4.2.2.1 Copiers. In copiers, two types of exposure are used. If the geometry of the photoconductor allows for flat-bed exposure, then it is possible to image the complete document by a single flash exposure. In early systems the photoconductor was supplied as a flat plate and the document could either be imaged through a lens arrangement, or, with the correct paper type and for documents which are only printed on one side, a direct contact technique was possible where the document is illuminated through the page. In more modern equipment, flash exposure can be very advantageous in achieving a high throughput speed which is important when considering large volume copying applications. Often in these systems, the photoconductor is made of a flexible material which is made into a belt. This allows the photoconductor to be held flat at the imaging/exposure station and then subsequently passed to the other processing stations. More commonly, the photoconductor is a drum which does not permit full-document, single-flash exposure. In these cases an optical arrangement of mirrors and lenses has been developed to scan the document and project the image on to the rotating drum. By careful mechanics, the focal distance is maintained throughout this process and minimal distortion of the image is achieved. By adjustment of the focal distance it is possible to photo-enlarge or reduce the document.

4.2.2.2 Laser printers. Laser printers, as the name suggests, use laser radiation to effect the discharge of the photoconductor. The earliest designs were developed for mainframe computer printing and were based on a standard copier engine where the lens/imaging system was replaced by a helium–neon laser (emitting red light at 633 nm). The laser light scanned the photoconductor surface and by switching the laser on and off, an image could be written on the drum surface. In modern printers, semiconductor lasers are used. These are small, very reliable, highly efficient and readily manufactured at low cost. Lasers emit specific wavelengths of radiation depending on the materials used in their construction: in laser printers the laser is often an infrared laser emitting at 780 nm. A rotating polygon mirror is used to direct the laser beam across the photoconductor surface and the high-speed switching of the semiconductor lasers allow a high-resolution pattern to be produced (Figure 4.2).

In the original laser printers, the laser radiation was used to discharge the photoconductor in a similar process to that in photocopying. Thus,

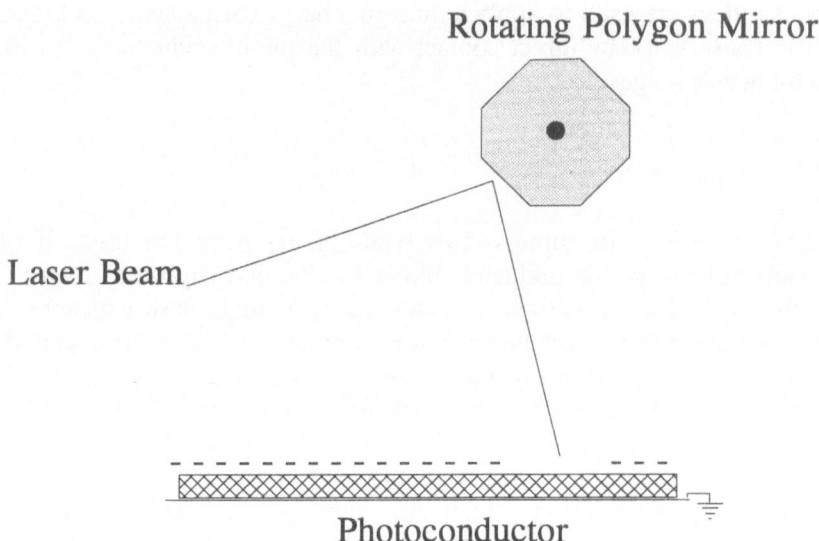

Figure 4.2 A polygon mirror directs the laser beam across the photoconductor.

when text information was being reproduced, a large percentage (*c*.95%) of the document is white background. This requires the laser to discharge a very large area of the photoconductor surface. Also, it is very important that pixels overlap slightly in order that complete discharge of the background occurs. It was realised that it would be faster and simpler if the process was reversed and that the laser addressed the areas where text is present. Thus, the image (black) areas are discharged and these are subsequently developed to produce the print. Failure to discharge the area between pixels is less important; also the process is much faster with text output since the laser is only discharging a relatively small area of the photoconductor (Figure 4.3).

4.2.3. *Image enhancement*

One problem with digital devices such as laser printers is that the images produced are made up of a series of pixels. These pixels are either in an 'off' or 'on' state and are addressed in a raster fashion. Each pixel is of a fixed geometry and position. In the case of office printers, pixel addressability is between 300 and 600 dpi (dots per inch). Increasing the number of pixels increases the resolution; however, more computing power is often required to process the information required to produce an image. Even with increased resolution, when trying to create a representation of certain shapes, a compromise must be reached.

As an example, consider a resolution of 10 × 10 pixels. In order to draw

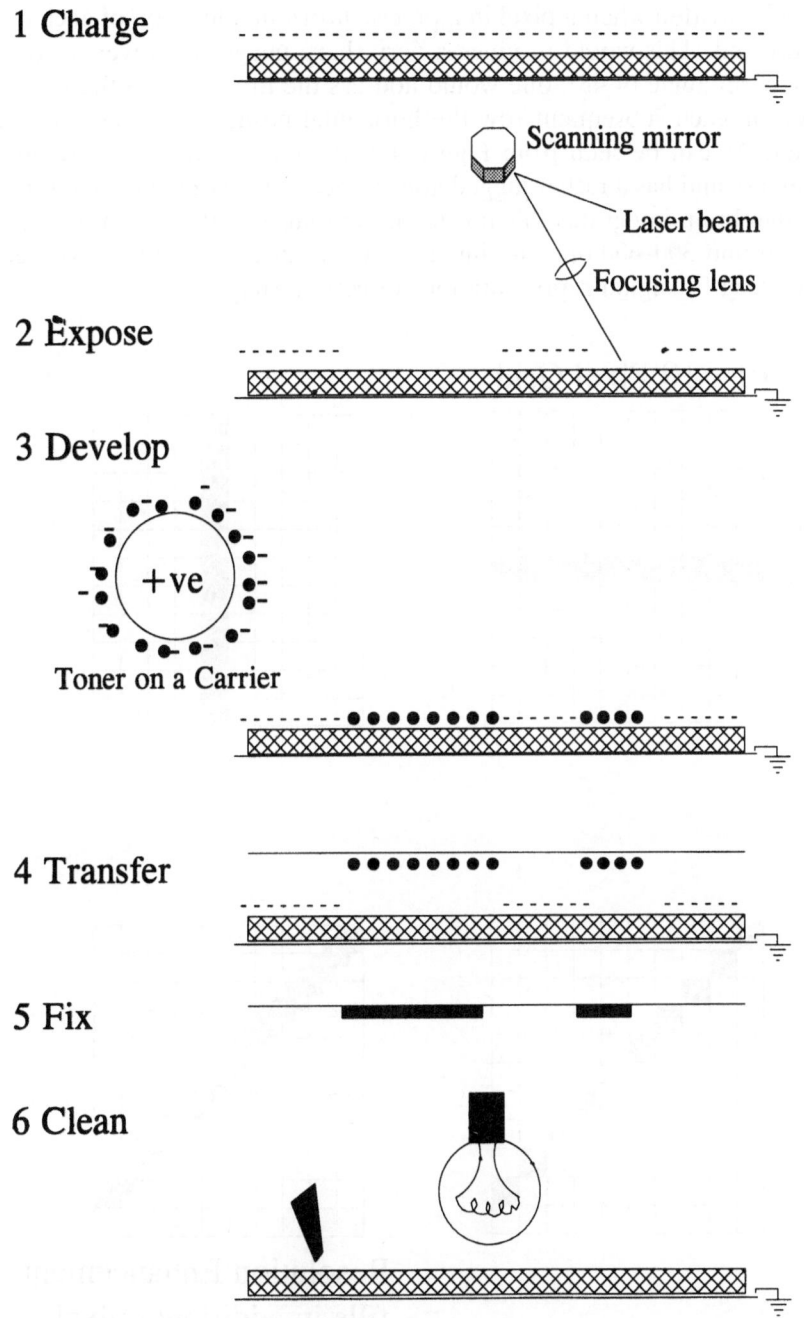

Figure 4.3 Reverse development: the electrophotographic process employed in modern laser printing.

a horizontal line one-pixel wide and ten-pixels long, a single row of pixels can be written. Similarly, a vertical line of the same thickness and length may be written when a pixel in a precise horizontal location of each row is addressed. This would produce a very sharp image. However, to draw a line at an angle of 45°, one would address the first pixel on the first row, then on each subsequent row the horizontal position is advanced by one pixel. As can be seen from Figure 4.4, the edge of the line resembles a staircase and has a rather jagged appearance. At very high resolutions, the human brain integrates this image and the jagged effect is not perceived. At around 300–600 dpi, the human eye can distinguish the jagged edges produced by digital representations of certain shapes.

10 x 10 Pixel Grid

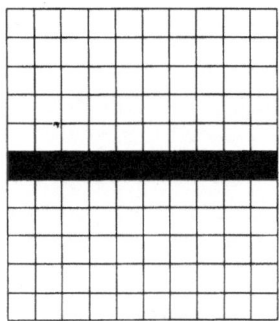 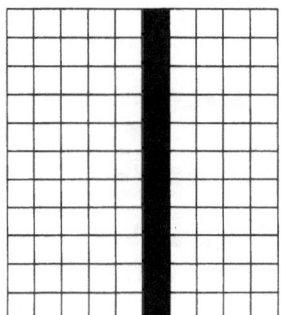

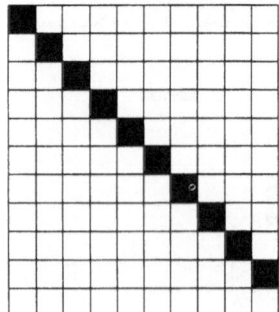 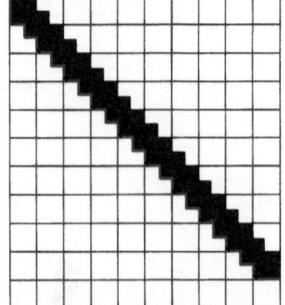

Resolution Enhancement
fills in additional pixels

Figure 4.4 Resolution enhancement: half-size pixels are used to fill in jagged edge on diagonal line.

In printers such as the HP Laserjet III, it is possible to vary the pixel size by sophisticated hardware control (Sanders, 1993). This enables the production of smaller pixels, which can be placed adjacent to large pixels and utilised to smooth-out unwanted jagged effects. The printer is able to do this by first preprogramming a series of bitmap tables which indicate patterns of pixels, where, by introducing smaller pixels, the desired effect is achieved. The printer's on-board computer analyses the bitmap pattern being processed, performs a pattern recognition sequence and, if an identified pattern is recognised, the additional small pixels are added to the sequence following preprogrammed rules.

4.2.4 Development

Once the electrostatic image of the document has been formed, this has to be translated into a physical effect if the image is to be visualised. Many techniques have been demonstrated (Thourson, 1972). The most common techniques used to prepare a 'hard copy' rely on the electrostatic attraction of fine-coloured powders to the image. This powder, referred to as the 'toner', is then used to form the copy or print.

Due to the nature of the electric field on the photoconductor, any toner sprinkled on the surface is attracted to areas where there is high field contrast. This means that areas where there is a transition from charged to uncharged states will have a stronger attraction for toners. When imaging a document, fine line detail will be reproduced easily by this process; however, large areas, where there is little field contrast, will show poor attraction for the toner and so may not develop properly. To correct for this, a development electrode is used. This is a device which, when connected to the earth contact of the photoconductor and positioned close to the photoconductor surface, maintains field contrast in large charged areas and so ensures even toner distribution.

Most toners rely on tribo-electric charging to ensure correct development of the electrostatic image. When two dissimilar materials are brought into frictional contact, an electrical charge of opposite polarity is developed on each. By careful selection of these materials it is possible to produce negative or positively charged toner. Matching this toner to the charge of the electrostatic image means that it is possible to choose whether the toner is attracted to the charged (normal) or uncharged areas (reversed devlopment).

4.2.4.1 Dry cascade development.

In this technique, the fine toner powder is mixed with large beaded material referred to as the 'carrier'. The carrier is often made of iron or iron oxides and is surface coated in order to modify its charging properties. This is then agitated with the toner powder, the friction producing electrostatic charge. As the carrier maintains an opposite charge to the toner, each carrier bead is coated with an even distribution of small toner particles. When the developer is allowed to cascade over the photoconductor surface, only in areas where sufficient field strength exist are the toner particles attracted away from the carrier bead. This ensures that any stray toner particles landing in areas of low field strength are re-attracted by the carrier beads (Figure 4.5).

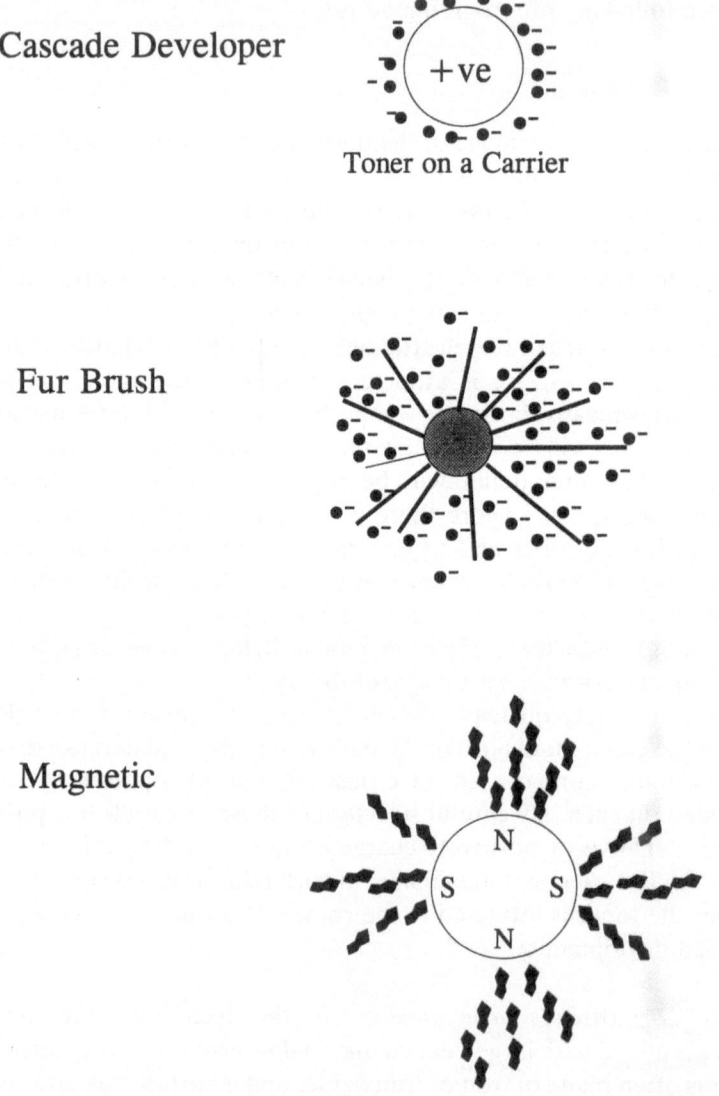

Cascade Developer

Toner on a Carrier

Fur Brush

Magnetic

Figure 4.5 Examples of dry cascade toner, fur brush and magnetic brush development.

4.2.4.2 Fur-brush development.   In the case of fur-brush development, the carrier is replaced by a fine fur brush (Bolton and Goetz, 1956). The friction between certain types of natural fur and the toner powder can be used to produce the correct polarity of charge. The fibres of the fur serve to charge the particles of toner and to hold the particles until contact with the photoconductor under the correct field conditions cause toner release. Natural fur fibres are subject to the effects of humidity and, as such, can present problems of control under varying environmental conditions.

4.2.4.3 Magnetic-brush development.   In this case the brush fibres are simulated by small ferromagnetic particles. The effect is similar to that observed when iron filings are attracted to a magnet. The particles align themselves with the direction of the magnetic field and appear to form fibre-like filaments. Friction between the toner and the coated ferro-magnetic particles causes tribo-charging and holds the toner until transfer to the photoreceptor (Young, 1957). This is the technique most widely used in commercial copiers and printers.

4.2.4.4 Liquid toners.   By dispersing the toner in an insulating hydro-carbon, the toner particles are charged during the dispersing process and the electrostatic charge helps maintain the dispersion (Metcalfe and Wright, 1956). The photoconductor can then be immersed in the liquid and the charged particles migrate on to the photoconductor by electrophoresis. This method of toning images was often employed in machines using zinc oxide photoconductive paper. It is less common now but liquid develop-ment systems are often employed in high-resolution applications such as colour proofing.

4.2.5 Fixing and cleaning

The fixing of the toner to the paper can be effected in three ways.

- Heat fusion is by far the most common fixing method. Most systems favour the use of a heated roller, normally a rubber roller impregnated with silicone oil which is preheated to 90°C. When the copier or printer is switched on, the waiting time is usually associated with the fusing stage warming up. The heated roller is designed to melt the toner on to the page, and the toner design, coupled with the treated rubber, ensure toner does not stick to the roller. Alternatively, in higher speed systems, flash fusion can be used. Special lamps with a specific heat output are used to rapidly heat the toner which then adheres to the paper.
- Cold fusion is a process where pressure is applied to the toner in order to make it fuse to the paper. The mechanical energy applied is designed to be high enough to effect the melting of the toner and so fuse it to the

paper. Generally this technique is less effective than heat fusion and can give rise to unacceptable shine on the toned areas.

● Solvent fusion is the least common technique. Solvent can be used to soften the toner and 'melt' the resin into the page. However, for normal resin systems, reasonably harsh solvents would be required to swell and soften the resin. In an office environment, control of solvent emission is very important and would put stringent requirements on the design of the print engine.

Cleaning of the photoconductor prior to re-use is an important last step. An erase lamp and/or a.c. corona is used to fully discharge the photoconductor. Once discharged, any toner which has not been transferred can be removed. Once cleaned and fully discharged, the photoconductor can be re-used.

4.3   Consumables

4.3.1   Photoconductors

In all systems the photoconductor has a finite life and as such can be classed as a consumable. The lifetime of photoconductors are often measured in thousands of copies and they will normally require replacing several times during the lifetime of the engine. In some cases, the photoreceptor may be designed to outlast the lifetime of the engine. Alternatively, in certain applications the photoconductor is used once only and is consumed during the copy process.

The photoconductor has advanced significantly since the days of Carlson's sulphur plates. Plates made from sulphur or anthracene were not sensitive or robust enough to form the basis of a good copy system. Amorphous selenium vacuum deposited on to a flat plate was used in the first commercial copier (Keck, 1952). Although high vacuum techniques were required in the manufacture of the photoconductor, the technology was able to be adapted to the mass production of photoconductors which in turn led to an explosion of office copiers. Selenium photoconductors were quite reliable and with the introduction of doping techniques many of the properties of the photoconductor could be controlled.

Other semi-conductor materials with photoconductive properties were investigated, such as cadmium sulphide. However, some of these inorganic materials are quite toxic and their use in an office environment may be undesirable. Zinc oxide photoconductors were also popular for some time. Zinc oxide could be formulated as a dispersion and coated on to a conductive substrate. Commonly, metallised paper was used. These photoconductors had poor lifetimes. However, since they were so cheap to manufacture, they were often not recycled and it was normal to fix the toner directly on to the photoconductor which became the copy. This type of process is rarely seen in office copiers, but is quite common for the production of print plates for short-tun to medium-run lithographic printing (see Chapter 2). More recently, amorphous silicon has been the focus of attention in the field of inorganic photoconductors. Although the technology to produce these materials is quite demanding, the resulting photoconductors are extremely durable and can have extremely long lifetimes.

To understand the design requirements of a photoconductor, it is necessary to appreciate the voltage changes that occur during the steps of the photoelectric process. Figure 4.6 illustrates a typical charge–discharge cycle for a photoconductor. The photoconductor passes below the corona wire and charge builds up on the surface. This charge must then be retained under dark conditions. On exposure to light, the photoconductor should then discharge. In a perfect world, all the charge would be dissipated by the light, but imperfections in the photoconductive materials often lead to small residual voltages being stored. These voltages will eventually dissipate; however, this can often take much longer than a single cycle of the machine. Some machines employ high-intensity erase lamps and an a.c. corona to neutralise any residual charge.

A perfect photoconductor would be one where the material charged rapidly to a high value, retained the charged in the dark and showed total discharge on exposure to low-intensity illumination. Furthermore, this performance would be maintained over the lifetime of the product.

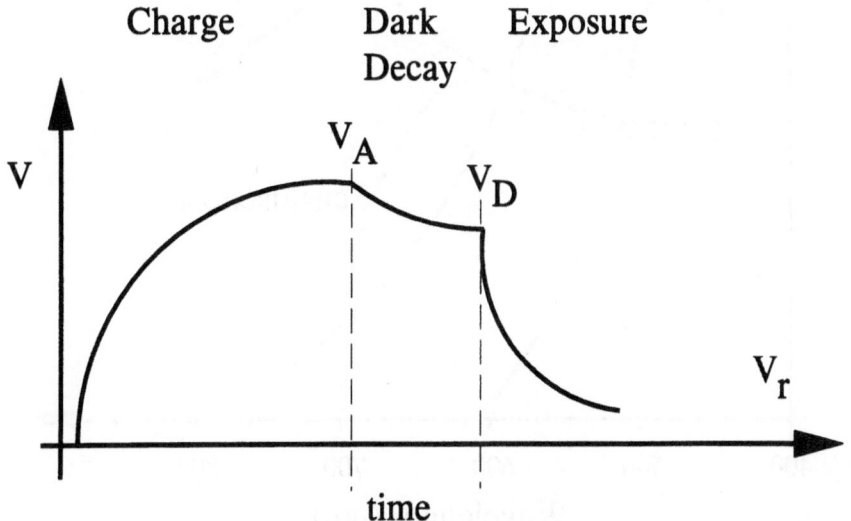

Figure 4.6 Charge–discharge cycle of a photoconductor during use. V_A = charge acceptance; V_D = dark voltage; V_r = residual voltage.

In the case of doped selenium, many of these criteria could be satisfied (Keck, 1952). The introduction of alloys could also extend the spectral response of the photoconductor. The zinc oxide photoconductors often contain sensitizing dyes which act to extend spectral response and increase light sensitivity (Young and Greig, 1954). Spectral response becomes important when imaging colour. Since not all documents being copied are black-and-white originals, the monochrome photocopier has to render a monochrome representation of the coloured original. Thus, the photo-conductor must show a uniform response to light intensity at wavelengths across the visible spectrum. A comparison of the spectral response of several photoconductors is shown in Figure 4.7

The high-vacuum techniques required for selenium manufacture and the sometimes limited spectral response of the photoconductors, especially at longer wavelengths, led researchers to consider organic materials. It was well known that the absorption characteristics of dyes and pigments could be readily tuned to meet specific requirements and this flexibility in organic materials would lend itself to the design of materials for specific laser applications.

The first commercially successful system was the polyvinylcarbazole

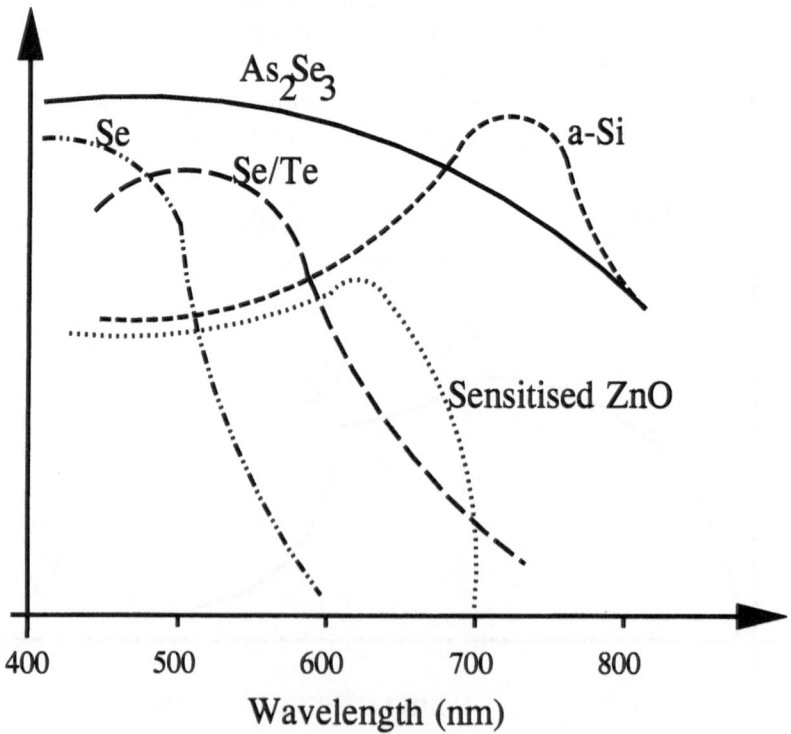

Figure 4.7 Spectral response of inorganic photoconductors.

(PVK) (1) based charge-transfer complex with trinitrofluorenone (2) developed by IBM (Schaffert, 1971). This near panchromatic photoconductor, although less sensitive than selenium, is easier to manufacture, the materials being solution coated on to the substrate.

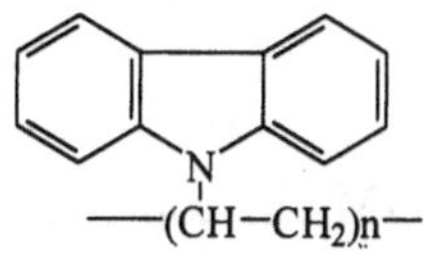

(1) 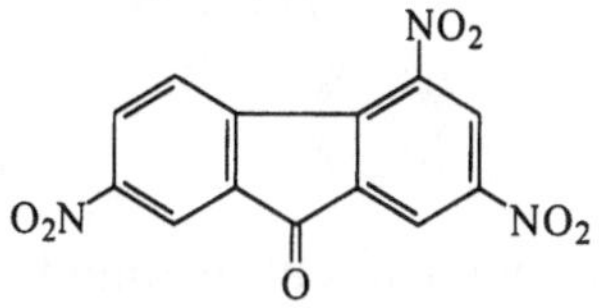   (2)

There now exists a wide range of organic photoconductive (OPC) materials and probably >80% of photoconductors now manufactured are organic. Both belts and drums are made from organic materials. The active photoconductive agents are normally formulated in resins and are designed to be both tough and flexible. Organic photoconductors differ from their inorganic counterparts in that they are often multilayered systems. The reason for this lies in an understanding of the photophysics of the photoconduction process.

When light strikes the photoconductor and an electric field is present, it is absorbed and acts to generate a charge pair. The transport or conduction of these charges enables the surface charge to be neutralised. In amorphous semiconductors like selenium, the selenium atoms are capable of generating a charge pair and transporting the charges. In the case of organic photoconductors, it is possible to separate these two processes and to develop materials which are optimised either for the ability to generate charge or to transport charge. These CGMs (charge-generation materials) and CTMs (charge-transport materials) are usually coated in a layered fashion.

4.3.2 Charge-generation materials (CGMs)

CGMs are often selected on the basis of spectral response. They are usually highly conjugated aromatic molecules, i.e. dyes and especially pigments. CGMs can be classified chemically.

4.3.2.1 Azos.

The most typical members of this class are the bis-azo pigments. These are often symmetrical. The central unit is usually the diazo component and is derived from readily available diamino compounds. Once diazotised, this is reacted with two equivalents of the coupling component. The coupling components are often amide derivatives of BON acids. The photoconductivity of azo compounds was first observed in 1969

(Rau, 1969). Chlorodiane Blue (Champ and Shattuck, 1975) (3) is one of the earliest examples to be utilised commercially.

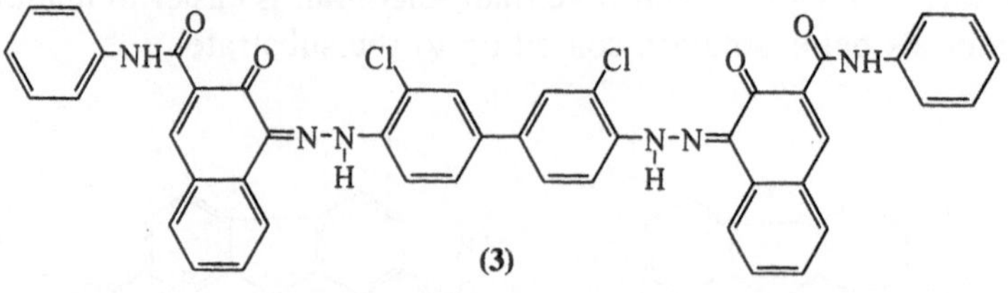

(3)

Many azo-pigments have found use as CGMs in copiers. The spectral sensitivity can readily be matched to white light copiers. Certain tris-azo pigments have also been used in laser printers. Examples of some of the most commonly used azos are shown in Figure 4.8.

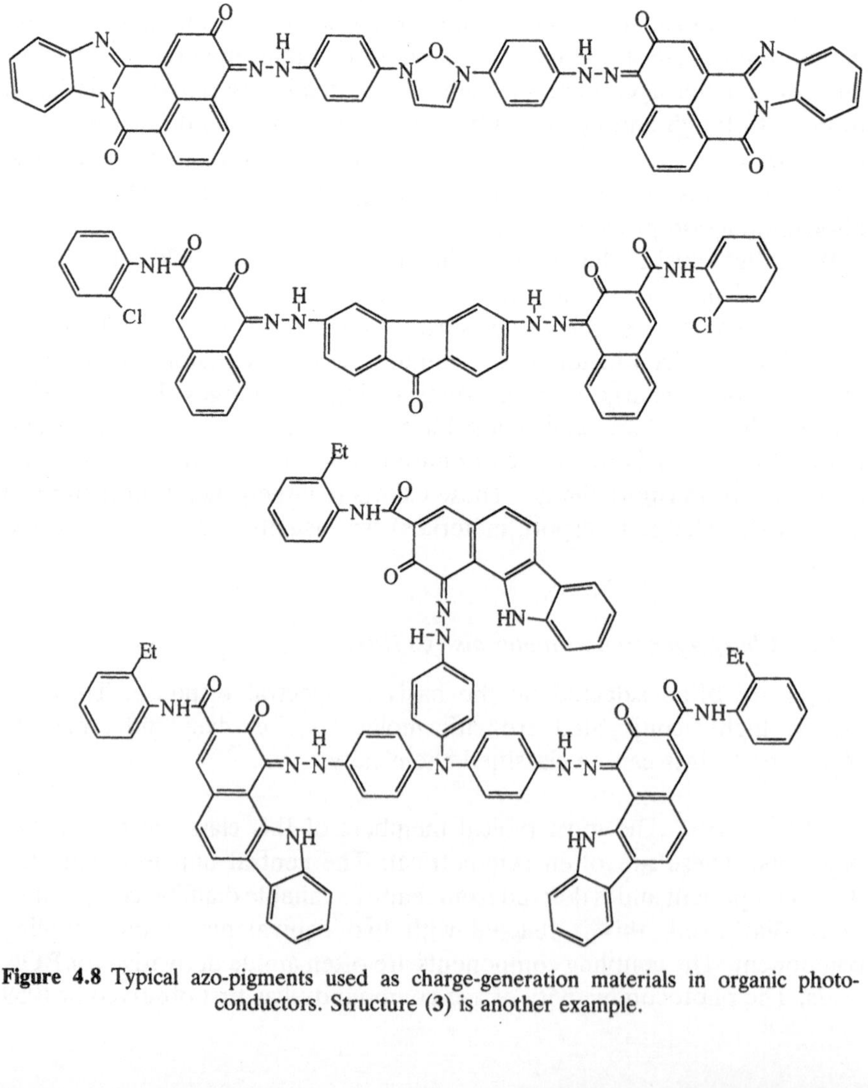

Figure 4.8 Typical azo-pigments used as charge-generation materials in organic photo-conductors. Structure (3) is another example.

4.3.2.2 Perylenes. Derived from perylene-3,4,9,10-tetracarboxylic acid, this class has received much attention in the literature since the early reports by Regensburger and Jakubowski (1972). The synthesis of these materials is reasonably easy; perylene-3,4,9,10-teracarboxylic dianhydride (**4**) is condensed with amines in solvent (Scheme 4.1). By varying the nature of the nitrogen substituents, the appearance of the pigment can be changed. Perylenes absorb mainly in the visible part of the spectrum and are really only of use as CGMs for visible-light applications, i.e. copiers.

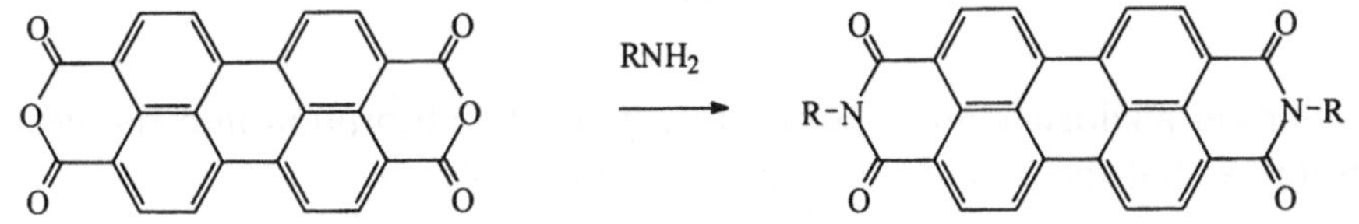

(**4**)

Scheme 4.1

4.3.2.3 Squaryliums. This class can be conveniently synthesised by a condensation reaction of squaric acid and substituted anilines (Scheme 4.2). Originally used as sensitisers in zinc oxide photoconductors (Kampler, 1970), they were found to show photoconductivity by Champ and Shattuck (1974). Squaraines show intense absorptions in the near infrared, and this makes them ideal for laser printer applications.

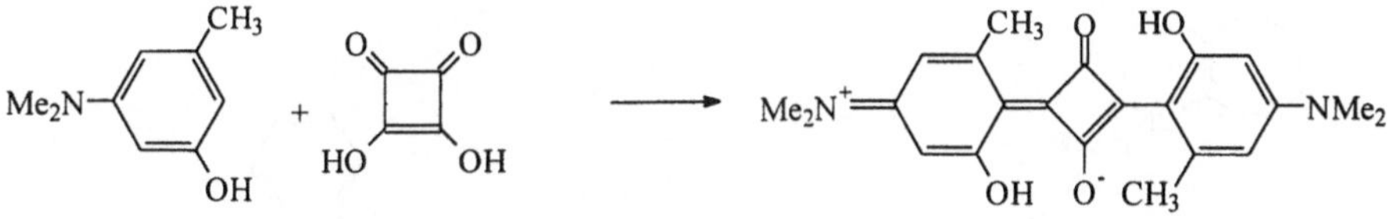

Scheme 4.2

4.3.2.4 Thiapyriliums. The sulphur analogue of pyrilium is prepared from the parent pyrilium by treatment with sodium hydrogensulphide. One of the simplest examples is prepared from the pyrilium which results from the condensation reaction of *p*-dimethylaminobenzaldehyde with two equivalents of acetophenone in perchloric acid (Scheme 4.3). The photoconductivity of this material was discovered by Light (1971) and has been the subject of several papers on the effects of 'dye aggregation' and benefits in photoconductivity (Dumage *et al.*, 1978). These materials have been utilised by Kodak in some of their white light copiers. The absorption characteristics of the simple thiapyriliums render them unsuitable for laser application. Extending the conjugation and making the selenium and

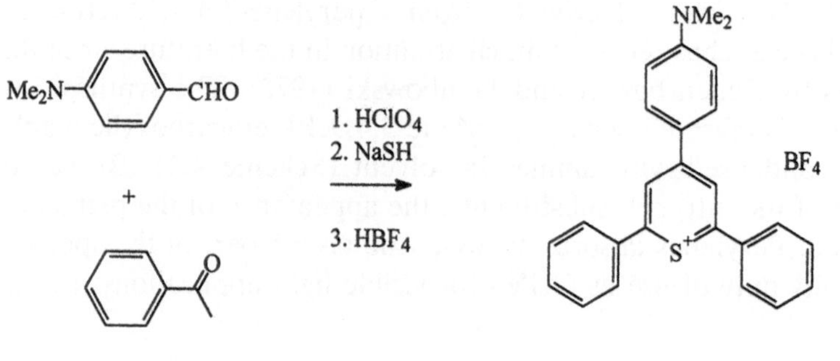

Scheme 4.3

tellurium analogues are reported to extend the absorption into the near infrared, but this has not been used commercially.

4.3.2.5 Phthalocyanines. Phthalocyanines are readily synthesised from either phthalonitrile, phthalic anhydride or diiminoisoindolines by the reaction with metal salts (Scheme 4.4). In this reaction, it is normal to isolate the metal phthalocyanine. However, when an alkali metal salt is used (such as sodium, lithium etc.), the metal can easily be removed by acid treatment to give the metal-free phthalocyanine (Scheme 4.4). Careful choice of the ring substituents can influence whether the resulting phthalocyanine is soluble or pigmentary. The central metal and ring substituents also effect the absorption characteristics. Many crystalline phthalocyanines exhibit polymorphism, whereby it is possible for the

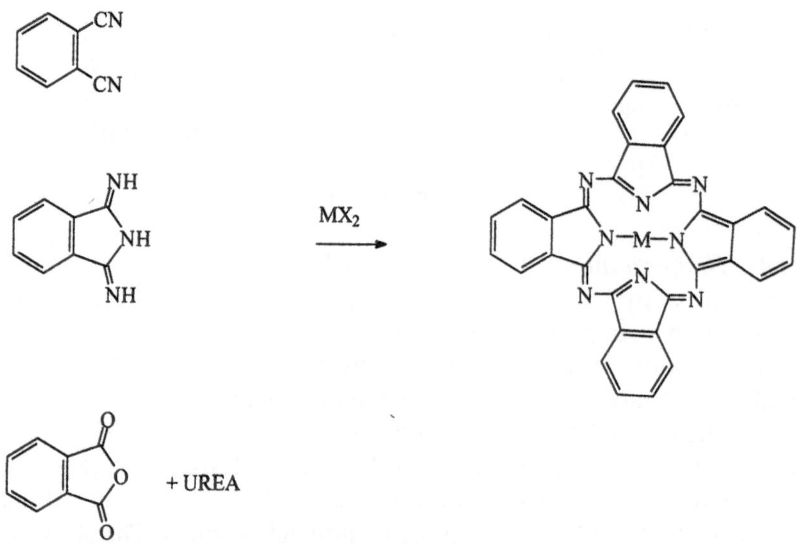

Scheme 4.4

molecules to stack in a variety of arrangements giving rise to different crystal structures. These differing crystal arrangements can affect the absorption spectrum of the phthalocyanine and have a significant effect on the photoconductive properties. There have been extensive studies of phthalocyanines and the associated polymorphism of particular molecular species. Phthalocyanines and the related naphthalocyanines absorb strongly in the red and near infrared areas of the electromagnetic spectrum. As such, they are particularly well suited to laser printer applications.

Metal-free phthalocyanine, in particular the x- and τ-forms, has been the subject of much study and was popular in laser printer applications (Enokida *et al.*, 1991). Copper phthalocyanine, in particular the ε-form, has been used in both OPC drum applications and lithoplate technology (Enokida and Hirohashi, 1992). Chloro-aluminium and chloro-indium phthalocyanines have been studied extensively (Kato *et al.*, 1986). Chloro-indium phthalocyanine is usually vacuum deposited and then exposed to solvent vapour to achieve maximum sensitivity. More recently, the patent and academic literature has been most active in the fields of titanyloxy, vanadyloxy, and chloro/hydroxy-gallium phthalocyanines (Enokida *et al.*, 1990). Studies, in general, have concentrated on new morphologies which are resulting in very high sensitivity in photoconductor applications. Of these 'new' materials, titanyloxyphthalocyanine is used commercially in certain Canon engines, and is also used in some of the Hewlett Packard LaserJet 4 series.

4.3.2.6 Other photoconductive materials. There are several other classes of materials which show photoconductive properties. The dithio-ketopyrrolopyroles (**5**) have been the subject of patent and academic literature (Mizuguchi and Rochat, 1988).

Similarly, the photoconductive properties of quinacridones (**6**) (Baranyi *et al.*, 1990a), thioindigos (**7**) (Baranyi *et al.*, 1990b), azulenium dyes (**8**) (Katagiri *et al.*, 1986) and other colorants have been reported. Commercial systems are currently favouring the more traditional classes. Perhaps one exception to this is the use of dibromoanthanthrone (**9**) in white light copier applications (Takei *et al.*, 1989).

4.3.3 Charge-transport materials (CTMs)

Charge-transport materials can be classified as either 'hole transport' or 'electron transport' according to the sign of charge being transported. By far the most common are hole-transport materials, with most commercial machines using these materials.

4.3.3.1 Hole transport. Hole-transport materials are, in general, materials which can readily give up electrons, i.e. they are easily oxidised.

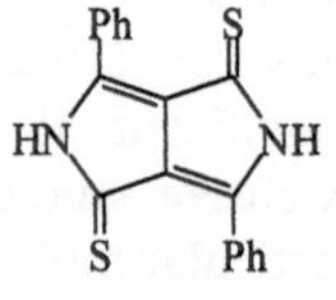

(5)

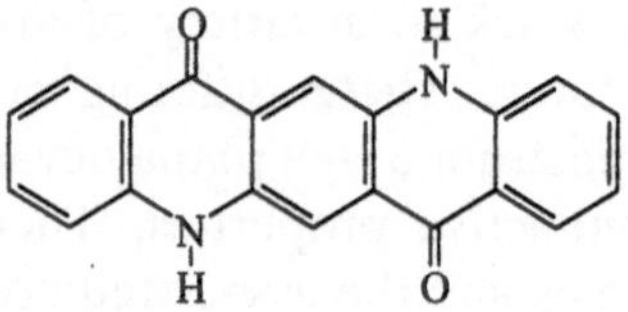

(6)

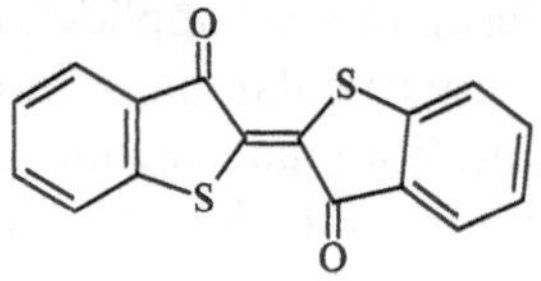

(7)

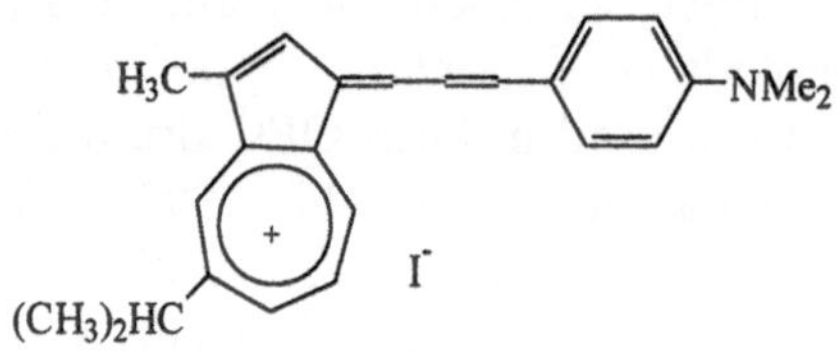

(8)

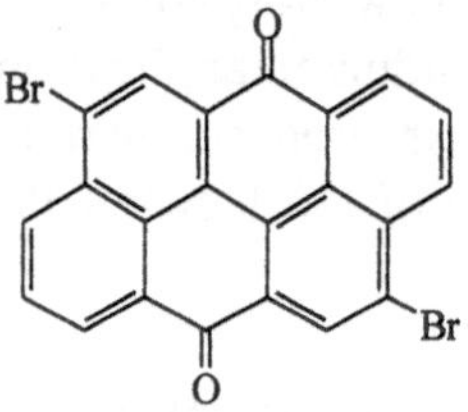

(9)

More often, they tend to be low-molecular-weight aromatic materials with tertiary amine substituents. Although this general description would tend to suggest that many species could be used as hole-transport materials, there are relatively few classes of materials which appear to exhibit the correct properties.

4.3.3.1.1 Triarylamines. Based on the triarylamine structure, tri-*p* tolylamine (**10**) could be considered the parent of this class. This class typically show high charge mobilities which makes these materials very effective as CTMs in high-speed applications. As such, these materials often figure in patents relating to high-speed devices and are commonly cited in combination with CGMs such as the titanyloxy- and hydroxygallium phthalocyanines.

Tri-*p*-tolylamine has been the subject of much study and has been used commercially (Eastman Kodak, 1972). One drawback with this material is

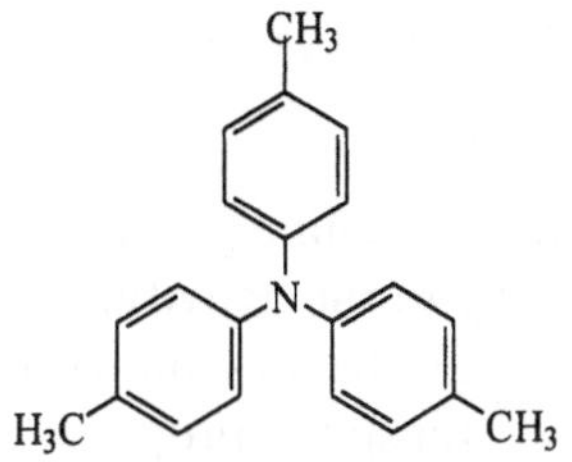

CH₃

(10)

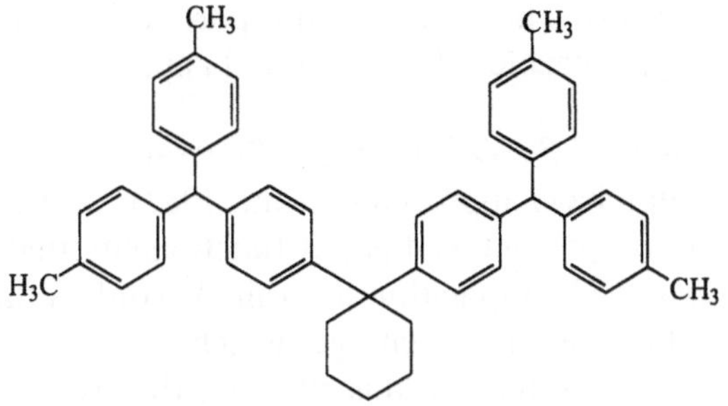

CH₃ CH₃

H₃C CH₃

(11)

CH₃ H₃C

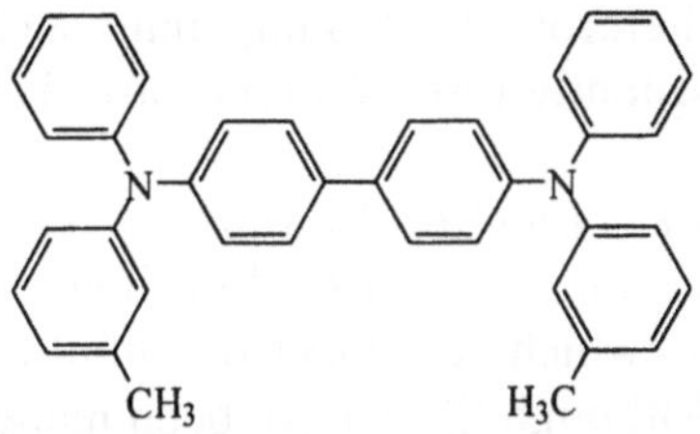

(12)

the poor compatability with conventional resins (polyester and bisphenol-A polycarbonates). At high concentrations, the material shows a tendency to crystallize in the polymer. By restricting the concentration of CTM used, the mobility is diminished and so the photoconductive properties are affected. Several patents have appeared concerning more elaborate arylamines; usually these involve the replacement of one of the tolyl groups with alternative aromatic systems such as pyrene (Kikuchi *et al.*, 1992).

Dimeric materials have similarly been studied. Kodak (11) and Xerox (12) both have claims to improved CTM systems which are based on dimeric materials (Borsenberger, 1990). The presence of two amine centres in a single CTM molecule may assist the transport of charge through the CTL (charge transport layer) and so improve mobility.

4.3.3.1.2 Hydrazones. A very popular class of CTM, there has been a significant number of patents on various structural types and mixtures. The parent of this group is often referred to as DEH (N,N-diethylamino-benzaldehyde diphenylhydrazone) (13), and is derived from the reaction of diphenylhydrazone and N,N-diethylaminobenzaldehyde. This material was first patented by IBM and has figured in many of their products (Anderson and Moore, 1977). The related naphthyl derivative (14) was adopted by Canon for their early laser engines. Several heterocyclic derivatives have also been reported, in particular derivatives based on N-ethylcarbazole (15) have found application in Ricoh printers (Weis, 1992).

4.3.3.1.3 Other types. Several systems which utilise extended conjugation have appeared (Sasaki, 1982). In these examples, amine groups are always present as it is this functionality that is the key to the activity. The extended conjugation most likely confers the correct electronics and good delocalisation of charge which confers the best transport properties. Typical systems in this class are the styrilics (16) and the butadienes (17) which have good mobility properties.

Of the other systems which have been reported, most are aromatic amines and they utilise heterocyclic linking units such as oxadiazoles. Examples of the most frequently reported types are shown in Figure 4.9.

4.3.3.2 Electron transport. Electron-transport materials are molecules which readily accept electrons. They are often highly conjugated, low-molecular-weight aromatics which have electron-withdrawing substituents, especially cyano groups. Although there have been many reports of active materials in the literature, to date there have been no examples of this technology in use in current commercial systems.

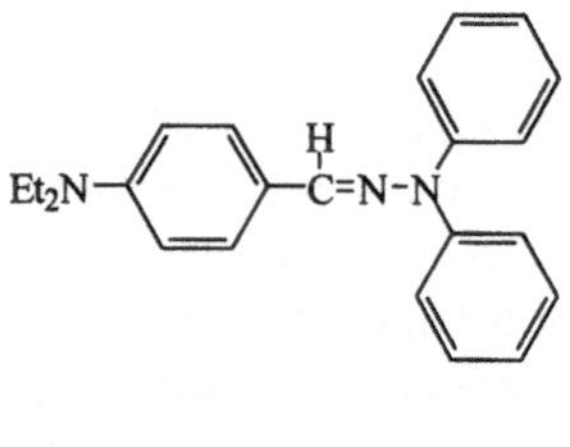

(13)

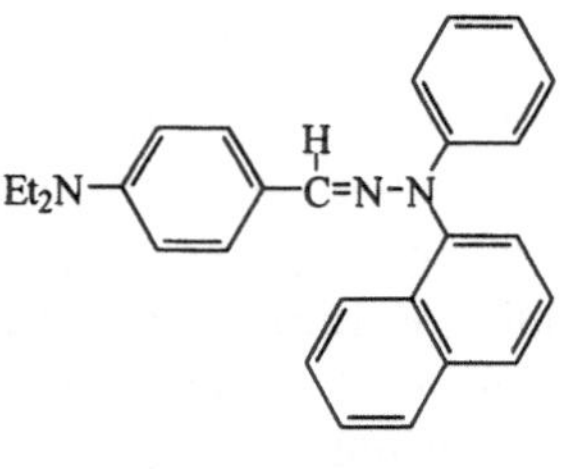

(14)

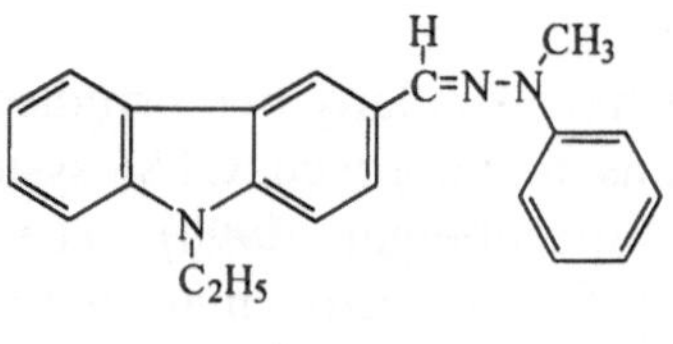

(15)

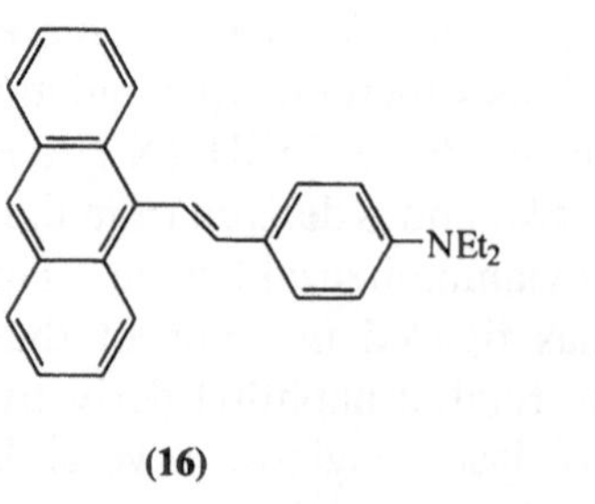

(16)

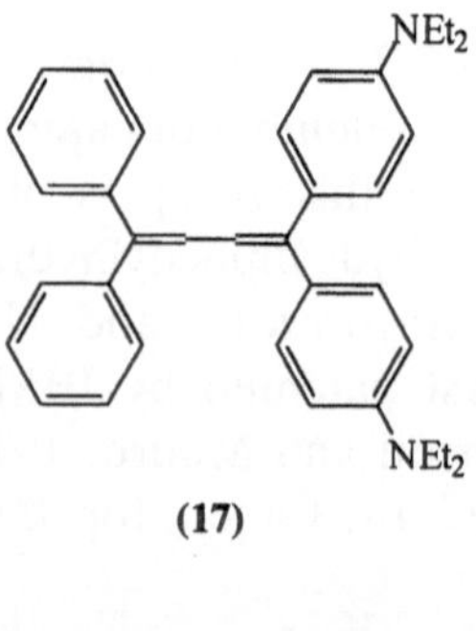

(17)

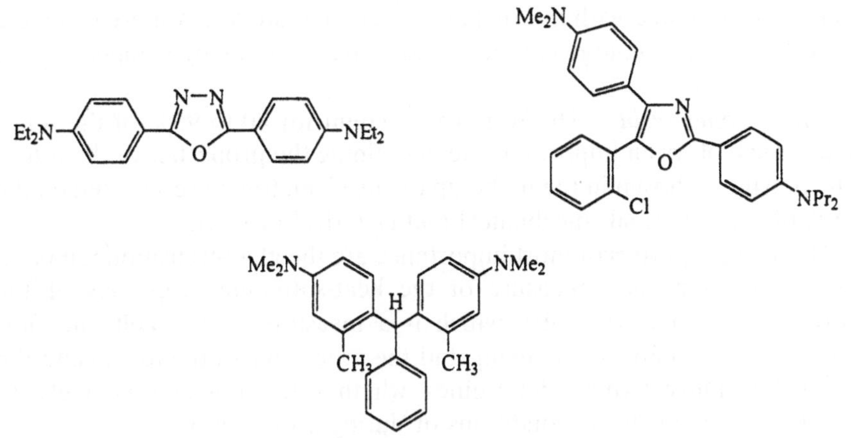

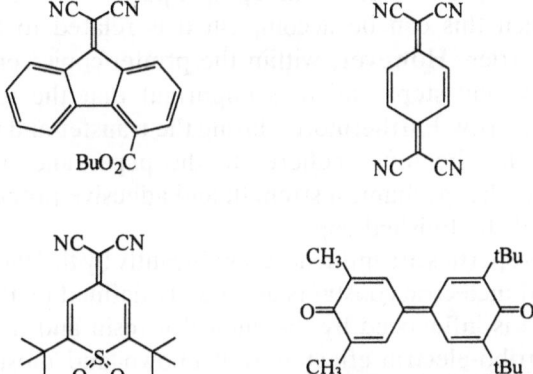

Figure 4.9 Examples of hole-transporting charge-transport materials. Structures (**16**) and (**17**) are other examples.

Figure 4.10 Examples of electron-transporting charge-transport materials.

Three main classes are most commonly reported in the literature (Ong *et al.*, 1983). The TCNE (tetracyanoethylene)//TCNQ (tetracyanoquino-dimethane) types are molecules which bear resemblance to tetracyano-ethylene or the related tetracyanoquinodimethane. Analogous to these are the sulphones, which show very similar conjugation patterns. Finally recent activity has concentrated on unsymmetrical quinones. Examples of these are shown in Figure 4.10.

4.3.4 Developer

Developer is the name given to the package used to form the image on the charged drum. In the case of liquid development this is a toner dispersed in an insulating liquid such as isopar. For dry-development systems, the developer can be just toner, in the case of single-component systems, or a mixture of toner and carrier. The toner is primarily made up of a resin which is formulated with internal and external additives. A more complete review of the technology can be found in publications by Cooper (1992).

4.3.4.1 Toner resin. The resin can account for 40 to 95% of the toner and as such plays an important role in defining the properties of the toner. The polymers chosen in the make-up of any given toner are selected on the basis of their thermal, mechanical and electrical properties.

Thermal properties of most importance are the glass-transition temperature (T_g), which is a measure of the heat-softening properties of the polymer, the melt viscosity (which is a measure of the melt and flow properties which affects the fixing and the processing of the toner) and the thermal stability (which determines whether the polymer is likely to decompose under fusing conditions or during processing).

Mechanical properties of importance are impact strength and adhesive/cohesive strength. In the production of most toner, it is a requirement to be able to grind the toner to get an acceptable particle-size distribution. The ease with which this can be accomplished is related to the mechanical strength properties. However, within the printer/copier engine there are numerous frictional steps and it is important that the toner retains its mechanical integrity. Furthermore, during the transfer and fusing stages, it is important that the toner adheres to the paper and not the transfer rollers. Finally, the mechanical strength and adhesive properties affect the substantivity of the finished copy.

Electrical properties are influenced significantly by the choice of polymer. Resistivity and dielectric constants are mainly defined by the resin. Tribo-electric charge is influenced by the choice of resin and the nature of the carrier. The tribo-electric effect is well known and most resins can be classified by their tendency to acquire a greater positive or negative charge when in frictional contact with a charging surface (i.e. a carrier). For the simplest resins, the series runs from positive to negative as follows: polymethyl methacrylate > polycarbonate > polyester > polystyrene > polyethylene > polytetrafluoroethylene.

The most commonly used resins in the toner industry are styrene-acrylics and styrene-butadienes, which account for the largest consumption. These resins can be easily tailored by adjustment of the monomer ratios, the polymerisation techniques employed and the molecular weight. Polyesters are most prevalent in full-colour applications as they have many desirable properties (pigment dispersibility being a key feature) and are again easily tailored by control of molecular weight and cross-linking. Many other resins are used often as modifying agents to assist in gaining performance advantages.

4.3.4.2 Colorants for toners. The market for toners is currently dominated by black. Most black toners rely on carbon black as the primary colorant. Occasionally, in colour copiers, the black toner is made from combinations of pigments and dyes. For coloured toners, most copiers/ printers use pigments to produce the yellow, magenta and cyan trichromat for full colour, or in specific mixtures to generate shades for spot colour applications (company logos, highlighting, etc.). Pigments are selected according to their resin compatibility/dispersibility, heat stability (required during processing) and light stability. Many classes of pigment are used:

- phthalocyanines, such as CI Pigment Blue 15:1 (blue/greens) (**18**) (Kiuchi and Maki, 1977).
- rhodamines, for example CI Pigment Violet 1 (**19**) (Higashida, *et al.*, 1980) and quinacridones, typically CI Pigment Red 122 (**20**) (Canon, 1982a) (magentas and reds);
- dichlorobenzidines, usually CI Pigment Yellow 12 (**21**) (Xerox, 1979) (yellows); and
- azos (used for a wide spectrum of colours).

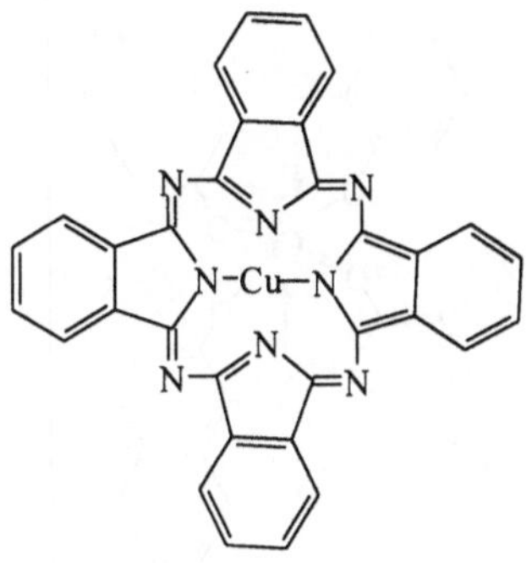

(**18**)

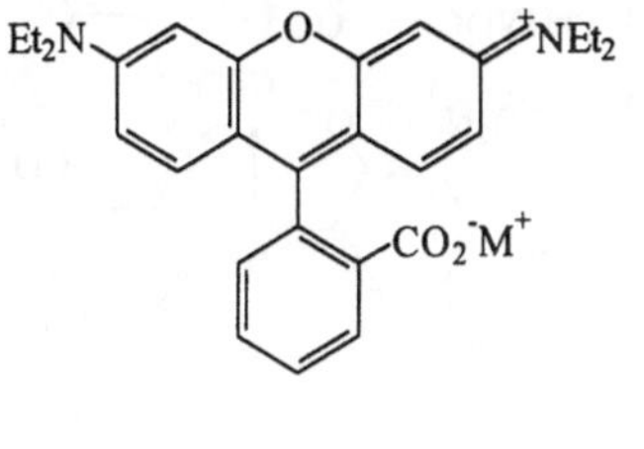

(**19**)

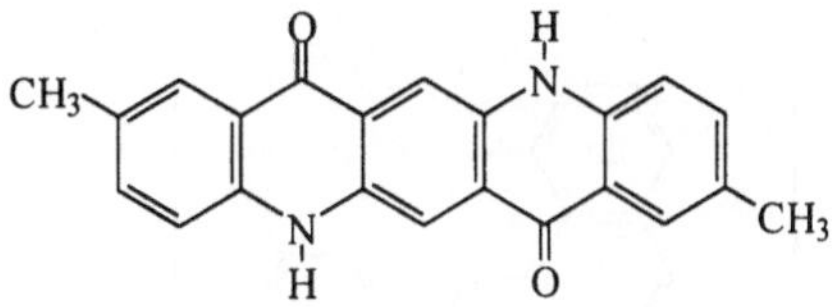

(**20**)

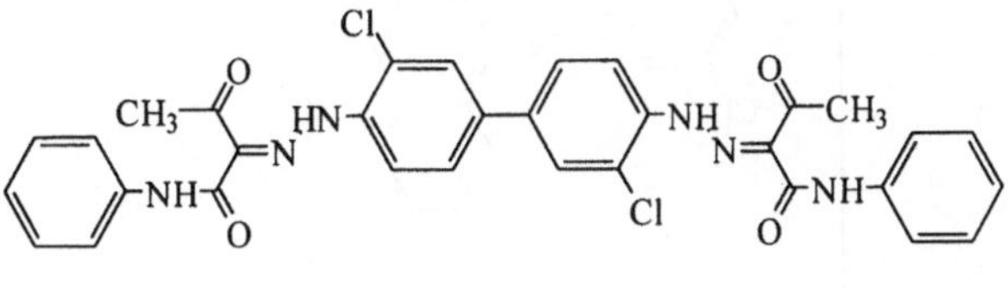

(**21**)

4.3.4.3 Charge-control agents. The electrostatic charge which is produced on a toner is directly related to the chemical composition of the toner. The Japanese were the first to introduce charge-control agents in toners. These agents affected the rate at which the toner charged and both the amount and sign of charge acquired. The net results was sharper, higher-density copies which could be more easily maintained over the lifetime of the developer.

The first charge-control agents were designed for negative charging toners. These were typically metal azo-complexes which were highly coloured and really only suited to black toner systems. Similarly, when the first positive systems were adopted, nigrosine dyes were used which again are highly coloured. More recently, with the growing emphasis on full-colour copying, colourless materials are actively being sought as it is often

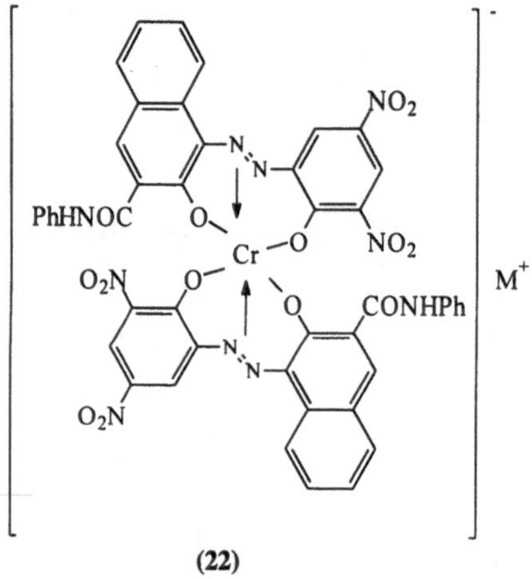

(22)

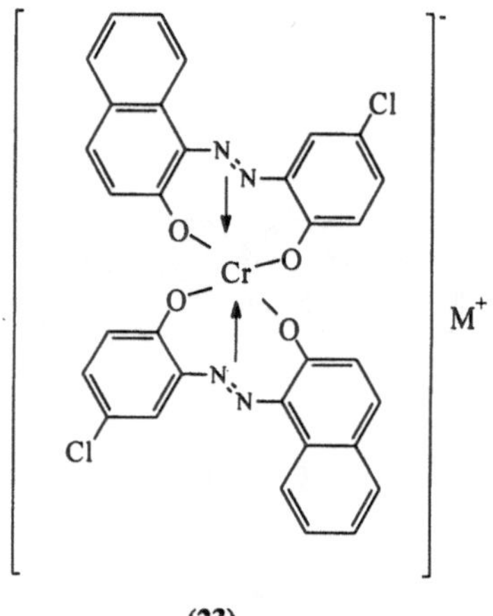

(23)

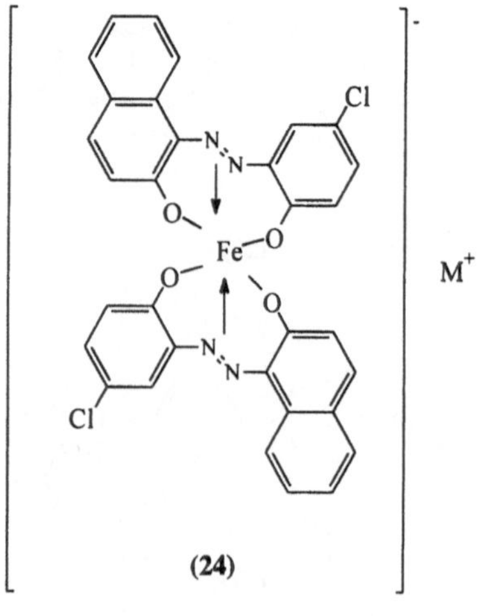

(24)

essential for all four colours (yellow, magenta, cyan and black) to be designed with the same charging profile.

Still the most common class utilised in negative CCAs is the azo–metal complex. These are usually 2:1 azo:metal complexes. The azo dyes favoured are normally derived from a phenol-containing diazo component and naphthol-derived coupling component. The position of hydroxyl groups with respect to the azo linkage is chosen so that they will be predisposed to providing effective coordination of a transition element, usually chromium(III). The structure of the final complex appears to be very important to the final properties. Electron substituents such as nitro groups are particularly beneficial but can convey undesirable effects such as thermal instability (important during processing) and have known mutagenicity (Ames positive). The most significant commercial materials derived from this class are Hodogaya's SPILON Black TRH (22), Orient Chemical Industries' S34, Zeneca's CCA7 and Hodogaya's T95 (23) and T77 (24) (Gregory, 1991).

Colourless negative CCAs are usually metal complexes or salts, but are normally formed from non-coloured precursors such as salicylates or the related BON acids. Again, chromium(III) is used as in BONTRON E81 (25) and E82 (26); however alternatives based on aluminium and zinc are available.

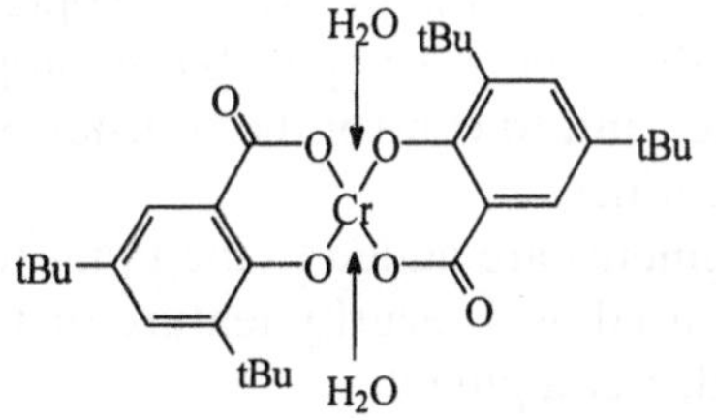

(25)

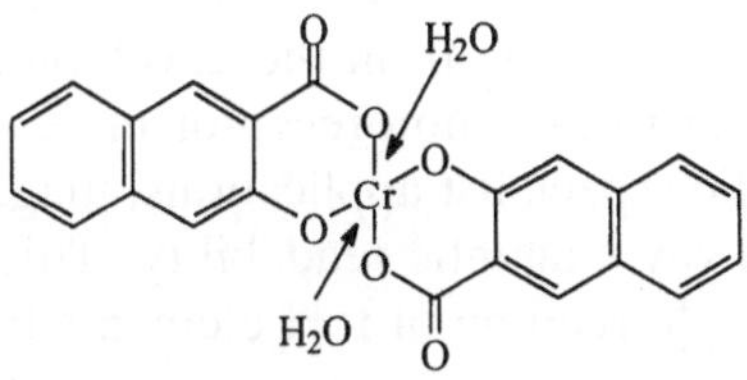

(26)

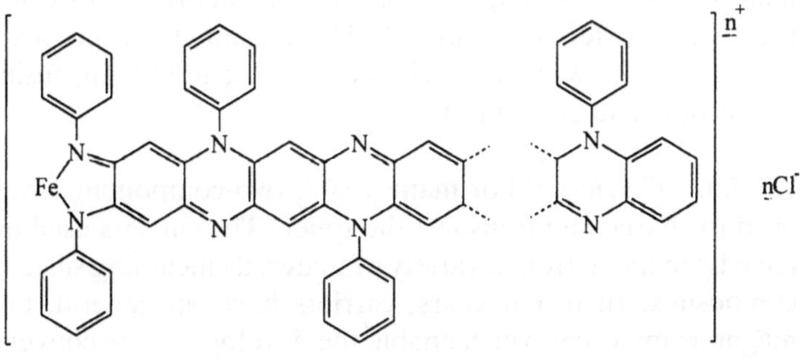

(27)

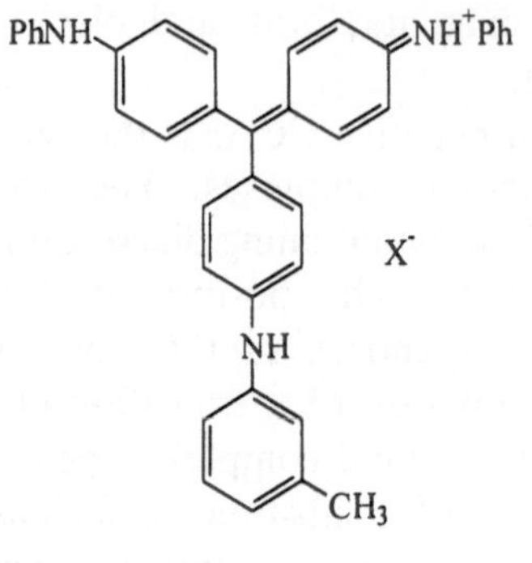

(28)

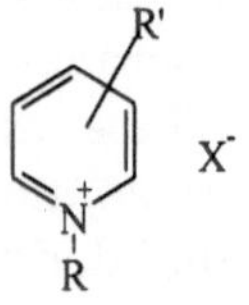

(29)

Positive CCA sales are still dominated by nigrosine, a cationic dye (**27**). As with negative CCAs, structures which can support delocalisation of the charge seem to exhibit the best properties. Triphenylmethane dyes (**28**) are also popular and are marketed by Hoechst. Colourless materials are in general limited to quarternary ammonium salts and related pyridinium salts (**29**). In general, these colourless materials do not perform as well as their coloured counterparts.

4.3.4.4. Magnetic pigments for toners.   These are most often added to toners as a means of controlling toner movement within the printer/copier engine. By using magnetic rollers within the engine it is possible to stop toner straying from the development stages and to control the transfer of toner to the photoreceptor in desired quantities.

For specialist applications, magnetic pigments are added to the toner to convey magnetic readability. This can be used as a security feature or to enable reading of text elements by specialist equipment.

4.3.4.5. Other toner additives.   There are many other agents which can, and are, often added to toners. These can either be added to the raw materials prior to compounding (internal additives) or can be added post-processing (external additives). The reasons for using many of these are numerous – flow control, charge control, modifying fusing and transfer properties, and cleaning properties.

4.3.4.6. Carriers.   For many years, two-component developer has been used to charge and transport the toner. The carriers used in the developer have been made from a variety of materials including silica, glass, steel and composites. In recent years, carriers have, in general, been made from magnetic materials which enable the developer to be conveyed by magnetic rollers.

Design parameters for carriers are concentrated on size, shape, surface area, electrical properties and magnetic properties. Carrier-size distribution is carefully controlled. Although smaller particles give better charging, these can more easily be carried out and can lead to photoreceptor damage since carriers are more abrasive than the toner. Irregular-shaped carrier gives better flow properties and can also give increased surface area which in turn gives better charging properties. However, smooth surfaces tend to be less abrasive and in general lead to longer carrier lifetimes.

Triboelectric properties of carriers are most important. Often carriers are surface treated to enable the correct triboelectric properties. Treatments can be simple heat processes which effect oxidation of the surface, or they can be chemical treatments with agents which modify the surface (e.g. silanes, polymers, etc.). Alternatively, ferrites can be prepared by varying their composition to achieve the same effect.

4.4 Advantages of electrophotography

Electrophotography is now a well-established technology. Photocopiers and laser printers are common equipment in most offices and with falling prices laser printers and personal copiers are now seen as affordable for the home-based office. There is a good sales and service infrastructure, and the technology is generally regarded to be reliable. Like all high-technology products, 'new innovations' appear at frequent intervals as manufacturers upgrade their product ranges; however, there is commitment to the installed base of products which is reassuring to the end user.

Electrophotography dominates the technology of photocopiers. Although alternatives have appeared (most notably ink jet, see Chapter 5), the speed, quality and ease of use of the electrophotographic copier has made it a formidable force in the market. From the single copy to medium-volume duplication, the process is reliable, rapid and economic.

Colour copying, although not a new technology, is growing in market importance. The multipass technology required to produce the copy has meant that colour electrophotographic copiers are more mechanically complex and are therefore more expensive. Also, the complexity of the paper-handling systems and the software required to render the colour images has naturally raised questions about reliability. However, many of these issues have been addressed and the cost of colour machines is now reducing dramatically. The speed of these systems is still a major advantage over the other technologies and even although relatively costly, large offices and copy shops tend to choose electrophotographic copiers over other types.

In the printer market, the laser printer is currently dominant in the office. This is primarily related to speed and ease of operation, but is equally related to quality of output. This is further reinforced by the software support which has developed around the laser printer industry. Hewlett Packard have led the way with the development of printer languages for their Laserjet printer series which have enabled very sophisticated control of computer output, In addition, the development of page description languages such as Postscript and the utilisation of the microprocessor power within the printers has brought desktop publishing into the office environment.

4.5 Limitations of electrophotography

Cost is probably the major limiting factor. Monochrome laser printers are still relatively expensive, and for most home computer users, and some small-business users, the initial outlay is prohibitive. As a result, the cheaper ink jet technology dominates the lower end of the market. The base price for the colour laser is yet to be established as this technology sector is still growing. Currently though, the newest printers are six times the price of similar quality ink jet systems and considerably more expensive than the cheapest colour ink jet printers.

Resolution and grey-scale reproduction are as yet still limitations to the technology. Resolution is increasing and although 600 dpi is standard, 1200 dpi is emerging and in some liquid toner systems, higher resolution has been shown to be achievable. When high-resolution detail and grey scale is required, conventional photography is probably the technology of choice (see Chapter 3). However, dye-diffusion thermal printing can give improved grey scale although resolution is still limited (see Chapter 6).

Volume of output is another important factor. The cost per copy is a factor which must be considered. Although this is fairly cheap for monochrome, when volumes become large enough, the cost of conventional printing press technology is extremely competitive (see Chapter 2).

4.6 Competition

The major rivals to the technology are ink jet for low-volume work and dye-diffusion thermal printing for full-colour graphics artwork.

Ink jet is a popular choice when cost is an issue. For instance, the lowest cost monochrome laser printer, the Hewlett Packard LaserJet 4L which operates at 300 dpi with resolution enhancement, has 26 scalable fonts and offers 4 ppm printing, is around £400. By comparison, the Hewlett Packard DeskJet 520 which operates at 600 × 300 dpi has four fonts and prints at 3 ppm is around £200 – half the price.

When colour is a consideration, the DeskJet 560 is probably one of the cheaper, quality printers and is currently around £300. This is a very attractive entry price for most people. Colour laser printing, by comparison, is still expensive. The QMS Colourscript 1000 currently retails for around £7000 and, although a Postscript printer, competes against the Deskjet 1200PS (also Postscript) which retails at £1200. Also, thermal dye-diffusion technology is a serious competitor, for example the QMS Colorscript 210 which retails for £4000 (*Computer Shopper*, February 1995, No. 84)

For monochrome, laser printers tend to offer higher resolution (commonly 600 dpi with image enhancement) at speed (at least 4 ppm), which is often the major advantage over the other technologies. Linked to this, the overall quality (fastness and look) of the laser output is superior and certainly in the office environment, is the usual choice. For the home market, cost is the most telling factor and, with the bonus of colour printing at less than the cost of the monochrome lasers, is the reason why a large percentage of ink jet printers are sold. In the office, the colour market is dominated by ink jet: whether the cost of lasers drops significantly is yet to be seen.

4.7 Synergy

There have been several technologies proposed which have similarities to electrophotography. Electrography (*Coates Toner News*, 1983), ionography (Buchan, 1987), lithography (Kagawa *et al.*, 1985), magnetography (Campbell Scott, 1988) and photoactive pigment electrophoresis (PAPE) (Tulagin, 1969) all have elements in common with electrophotographic printers. A newer technology, TonerJet (Johnson and Larsson, 1993), has more in common with ink jet technology (see section 8.5.3).

4.7.1 Photoactive pigment electrophoresis (PAPE)

The PAPE process has been the subject of much study. It is inherently simple in design and can theoretically be used most efficiently in single-pass colour applications. The process relies on the principle of electrophoresis – charged particles will migrate in the presence of an electric field. If the charged particles are photoconductive, then it is possible to effect a change in the charge on the particle by the action of light and so control the direction of migration.

The PAPE device is constructed with two electrodes, one of which is transparent. The photoconductive particles are charged and suspended in a fluid between the electrodes. When a field is applied, the particles collect on one elctrode. If an image is shone through the transparent electrode, the light will effect a sign change on the exposed particles which then migrate to the other electrode. If the photoconductors are designed to be sensitive to different frequencies of illumination then controlled migration of coloured particles can be achieved (Figure 4.11).

4.7.2 Electrography and ionography

Both these techniques make use of a dielectric material on to which the charge pattern can be written directly. In the case of electrography, a stylus is used to apply high-voltage charges to special papers or film-based substrates. Then the toner is applied directly to the substrate prior to fixing. Ionography utilises ion-deposition technology in which a special head sprays ions on to the dielectric surface. Most commonly, a smooth anodised aluminium drum is used. The toner is then applied to the drum and can be transferred and fused directly to the substrate by cold fusion. Ion-deposition printers are fairly rugged and are capable of high speed. Both electrography and ionography are discussed in more detail in Chapter 8.

4.7.3 Lithography

Conventional lithography is still one of the most common techniques used in modern printing (see Chapter 2). It relies on a photographic negative being used to produce an image on a metallic plate coated with an ultraviolet curable resin. If the imaging process utilises ultraviolet light, then where the resin is exposed it becomes cross-linked and cannot be easily removed. The plate is 'developed' by etching away the non-exposed resin. The resulting plate can be used in offset printing. Two variants have emerged which utilise elements of the electrophotographic process to remove the need for conventional silver halide photographic negatives. The first system is based on photoconductive lithographic plates. A photoconductor is introduced into the resin coating of the print plate. This allows the plate to be charged, imaged, either by white light or digitally by laser, and toned on to directly. The toner, once fused to the lithographic plate, is resistant to the etching solution, enabling the plate to be used in the conventional manner.

The second system replaces the photographic exposure negative with film which is developed by an electrophotographic variant. The film is made of a heat-deformable plastic which contains a coating of finely dispersed photoconductor particles. In the unexposed condition, the film is non-transparent to light. When the film is charged and exposed to light, the photoconductive particles can be made to change charge sign. Heating the film allows the charged particles to migrate. The migration created causes the film to become transparent and so creates a photographic image. This image is used in the conventional sense with ultraviolet curable plates. The advantage of this system is that the 'wet chemistry' is removed and that it is

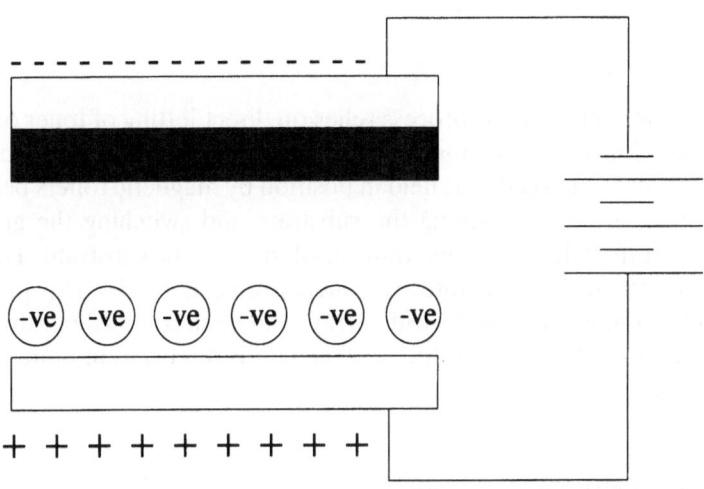

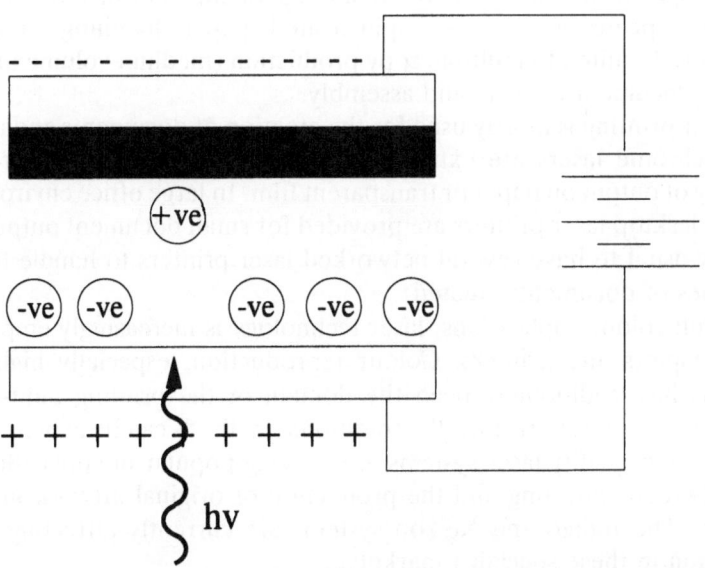

Figure 4.11 The PAPE process: light alters the charge on the pigment which then migrates to the opposite electrode.

possible to use digital techniques to produce the image since the film can be easily designed to be laser addressable. This technology is discussed in detail in Chapter 8.

4.7.4   TonerJet

As the name suggests, this process relies on direct jetting of toner on to the substrate. Conventional monocomponent magnetic toner is used. Once triboelectrically charged, it is held in position by magnetic rollers behind an electrostatic grid. By charging the substrate and switching the grid, it is possible to effect the controlled transfer of toner to the substrate. The toner is then fused to the substrate by conventional methods. The process is inherently simple and if page-wide arrays can be accomplished, this would ensure a reasonable printing speed. TonerJet is discussed in more detail in Chapter 8.

4.8   Uses of electrophotography

Photocopying probably still accounts for the major activity. The production of a single, either paper or transparent film, copy of an original document or artwork is one of the most common tasks addressed. For more sophisticated tasks, there are a range of copiers which provide faster process speeds and more sophisticated paper handling. These are particularly suited to multiple copy production (medium volume print run) or full-document copying and assembly.

Laser printing is mostly used for the creation of documents and artwork. Monochrome lasers are extremely common and are used to produce a variety of output on paper or transparent film. In large office environments, small desktop laser printers are provided for small document output, but it is now usual to have several networked laser printers to handle the large volumes of documents created.

In full-colour applications, laser technology is increasingly employed in both copiers and printers. Colour reproduction, especially high-quality colour, has traditionally been the domain of the printing industry. For single prints and short runs, the traditional methods can be expensive. As a result, high-quality laser systems are proving popular in applications such as full-colour proofing and the production of original artwork in graphic design. The Indigo and Xeikon systems are currently attracting a lot of attention in these specialist markets.

Interest in using the technology to print on to alternative substrates has been the focus of some studies. In particular, the Georgia Institute of Technology has been carrying out research into the application of the xerographic technique to the printing of fabrics.

4.9 Toxicology and the environment

As in all areas of technology, respect for the environment is an important issue. For many years legislation has existed to control the use and emission of toxic and harmful materials during production, and has sought to control the entry of harmful materials into the environment. Inorganic photoconductors which rely on toxic materials have certainly become less popular, the issues around disposal and recycling being a contributing factor. Organic materials must undergo stringent toxicology testing before they can be used and are thought to be less hazardous. Materials with poor toxicological profiles, such as the polyvinyl carbazole/trinitrofluorenone complex, are no longer used.

In recent years, concern on emissions of halogenated hydrocarbons during production processes has affected the design of new photoconductor manufacturing facilities. The selection of appropriate cleaning and degreasing agents and their containment is important. In the coating plants, solvent emissions are strictly controlled by legislation. Poly-carbonate resins, a particular favourite with the industry, only form stable solutions in a limited number of solvents, chlorinated hydrocarbons being the most common. Emphasis in resin design is trying to address this issue.

Another major concern is the introduction of heavy metals into the environment. These concerns centre mainly on the use of transition-metal complexes in toners, both in charge control agents, and, to a lesser extent, on the use of metal phthalocyanines in the photoconductors. Legislation in the state of California is often seen as a driving force in this field, and many manufacturers are looking at alternatives to chromium-based charge-control agents.

As with all industry, a greater emphasis is placed on recycling. The world's resources are finite and many companies now view it their duty to their customers base to offer recycling programmes. Photoreceptors are often sold as cartridges made of metal and plastic parts. When the toner supply is exhausted or the photoconductor worn out, these cartridges can often be returned to the manufacturer for recycling. In some instances, the cartridge is refurbished and re-used; alternatively, some components are re-used and others sent for reclamation.

As with all printing technologies, paper is a very important commodity. Copiers and laser printers use particular grades of paper which have been designed to be completely compatible with the electrophotographic process and this, to some extent, has precluded the use of recycled paper in the industry. However, the high-grade paper used by photocopiers is an excellent source of paper for the recycling process. The de-inking of paper is one of the most important factors in the recycling industry. Removal of toner residue is a prime concern. With this in mind, toner designers are pioneering new formulations which, while still giving durable print (necessary for archive purposes), can be more easily de-inked during recycling.

4.10 The future

As the technology is advancing several trends are emerging for the future. At the lower end of the market, the home-printer market cost floor has not yet been reached and there is a battle for market share between laser and ink jet printers. Ink jet has a distinct advantage in that it will always offer cheaper entry into full colour and is able to offer smaller, lightweight equipment, especially important in the portable market. However, there is likely to be a size reduction in laser printers. As colour dictates a neccessity for smaller photoconductors to enable compact multipass engines, monochrome printers will benefit from the spin-off technology.

In general, a higher degree of functionality can be expected from printers of the future. The distinction between the fax, the scanner and the copier will be lost as increasingly these become combined into one central instrument. As this becomes prevalent, the analogue copier will become less in evidence as digital processing becomes the norm.

Colour will be the area where the most significant changes will occur in the next few years. Since the introduction of the first digital copiers, there has already been significant advances in the hardware. This has enabled printers to break the £10 000 barrier and as the costs reduce, the use will increase. Primarily, the office-document market is still orientated towards the production of predominantly black-and-white documents. With few coloured originals, the need for colour copying has been limited. As the ability to produce coloured originals easily and, more importantly, with quicker printing speeds, the volume of colour documents which will be processed is going to increase significantly.

References

Anderson, H.W. and Moore, M.T. (1977) US Patent 4 150 987.
Arishima, K., Okada, T., Tate, A. and Hiroaki, H. (1984) US Patent 4 426 434.
Baranyi, G., Hor, A.M. and Loutfy, R.O. (1990a) US Patent 4 952 472.
Baranyi, G., Hor, A.M. and Loutfy, R.O. (1990b) US Patent 4 952 471.
Bolton, W.D. and Goetz, W.E. (1956) *Photographic Engineering*, 7, 137.
Borsenberger, P.M. (1990) *Journal of Applied Physics*, **68**, 5682.
Buchan, R. (1987) *Computer Technology Review*, 96–98.
Campbell Scott, J. (1988). In *Output Hardcopy Devices*, R.C. Durbeck and S. Sherr (eds), pp. 263–275, Academic Press, Boston.
Canon (1973) Japanese Patent 83043736.
Canon (1982a) Japanese Patent 58189652A.
Canon (1982b) Japanese Patent 58187949A.

Carlson, C.F. US Patents (1938) 2 221 776, (1939) 2 297 691, (1940) 2 357 809.
Champ, R.B. and Shattuck, M.D. (1974) US Patent 3 824 099.
Champ, R.B. and Shattuck, M.D. (1975) US Patent 3 898 084.
Coates Toner News, (1983) **10**.
Cooper, J.F. (1992) *An Introduction to Dry Toner Technology*, Toner Research Services.
Daimon, K., Nukada, K., Sakaguchi, Y and Jgarashi, R. (1994) *IS&T's Tenth International Congress in Non-Impact Printing Technologies*, p. 215.
Dessauer, J.H., Mott, G.R. and Bogdonoff, H. (1955) *Photographic Engineering*, 6, 250.
Dumage, W.L., Light, W.A., Marino, S.J., Salzberg, C.D., Smith, D.L. and Staudenmayer, W.J. (1978) *Journal of Applied Physics, 49*, 5543.
Eastman Kodak (1972) US Patent 3 706 554.
Enokida, T. and Hirohashi, R.J. (1992) *Journal of Imaging Science*, 36, 135.
Enokida, T. and Hirohashi, R.J. and Nakamura, T. (1990) *Journal of Imaging Science*, 34, 234.
Enokida, T. and Hirohashi, R.J. and Mizukami, J. (1991) *Journal of Imaging Science*, 35, 235.
Giaimo, E.C. (1957) US Patent 2 786 440.
Goldman, F. (1937) British Patent 464 122.
Gregory, P. (1991) *High-Technology Applications of Organic Colorants*. Plenum Press, London.
Griffiths, C.H. and Melnyk, A.R. (1982) US Patent 4 410 616.
Hana, P.S. (1920) Netherlands Patent 5142.
Hauser, O.G. and Ruckdesche, F.R. (1982) US Patent 4 395 471A.
Higashida, O., Moribey, I. and Yamamoto, H. (1980) German Patent 3 120 542.
Johnson, J. and Larsson, O. (1993) *IS&T's Ninth International Congress in Non-Impact Printing Technologies/Japan Hardcopy '93*, p.509.
Kagawa, Y., Ohishi, C., Nakao, S. and Asao, Y. (1985) US Patent 4 673 627.
Kampfer, H. (1970) US Patent 3 361 270.
Kaprelian, E.K. (1950) *Photographic Engineering*, 1, 53.
Katagiri, K., Oguchi, Y., and Takasu, Y. (1986) *Nippon Kagaku Kaishi*, 387.
Kato, M., Nishioka, Y. Kaifu, K., Kawamura, K. and Ohno, S. (1985) *Applied Physics Letters*, 46, 196.
Kato, M., Nishioka, Y. and Kaifu, K. (1986) US Patent 4 587 188.
Keck, P.H. (1952) *Journal of Optical Society of America*, 42, 221.
Kiuchi, M. and Maki, I. (1977) German Patent 2 815 857.
Kikuchi, T., Kanemaru, T., Senoo, A and Yashiro, R. (1992) US Patent 5 079 118.
Light, W.A. (1971) US Patent 3 615 414.
Loutfy, R.O., Hor, A.M., Hsiao, C.K. and DiPaola-Baranyi, G. (1985) *Journal of Imaging Science*, 29, 148.
Loutfy, R.O., Hor, A.M., and Rucklidge, A.J. (1987) *Journal of Imaging Science*, 31, 31.
Loutfy, R.O., Hor, A.M., Hsiao, C.K., DiPaola-Baranyi, G. and Kazmaier, P.M. (1988) *Pure and Applied Chemistry*, 60, 1047.
Metcalf, K.A. and Wright R.J. (1956a) *Journal of Scientific Instruments*, 33, 194.
Metcalf, K.A. and Wright R.J. (1956b) *Journal of Oil and Colour Chemistry Association*, 39, 845.
Mizuguchi, J. and Rochat, A.C. (1988), *Journal of Imaging Science*, 32, 135.
Murayama, T., Otsuka, S. and Tajima, T. (1979) US Patent 4 278 747.
Murayama, T., Otsuka, S., Tajima, T. and Sato, Y. (1981) US Patent 4 407 919.
Nields, M. (1988) *Mini-Micro Systems*, 59–60.
Ong, B.S. Keoshkeria, B. and McAneney, T.B. (1983) US Patent 4 474 865.
Patel, J. and Shimazu, K.I. (1981) British Patent 2 103 818.
Potter, J. (1987) *Computer Technology Review*, 102–104.
Rau, H. (1969), *Ber. Bunsen-Ges. Phys. Chem.*, 73, 810.
Regensburger, R.J. and Jakubowski, J.J. (1975) US Patent 3 904 407.
Sanders, J.R., (1993) *Fundamentals of Image Enhancement Technology, IMI Fourth Annual Laser Printing Conference*, July 1993.
Sasaki, M. (1982) British Patent 2 131 023.
Schaffert, R.M. (1971) *IBM Journal of Research and Development*, 15, 75.

Schaffert, R.M. (1975) *Electrophotography* (2nd edn) Focal Press, London.
Schieschek, U. and Klutke, F. (1939) German Patent 6 84 619.
Scozzafava, M., Chen, C.H. and Reynolds, G.A. (1985) US Patent 4 514 481.
Straughan, V.E. and Mayer, E.F. (1958) *Proceedings of the National Electronic Conference*, **13**, 959.
Takei, Y., Kijima. E., Goto, S. and Hiroyuki, N. (1989) US Patent 4 835 080.
Thourson, T.L. (1972) *IEEE Transactions on Electron Devices*, **4**, 495.
Tulagin, V. (1969) *Journal of Optical Society of America*, **59**, 328.
von Bronk, O. (1922) British Patent 1 88 030.
Weis, D.S. (1992) *Trends in Organic Photoreceptor Technology*, Diamond Research Corporation, The R&R Show, September 1992.
Xerox (1968) US Patent 3 383 993.
Xerox (1971) US Patent 3 553 093.
Xerox (1979) British Patent 2 059 618B.
Yamaguchi, Y., Tanaka, H. and Yokoyama, M. (1990) *Journal of the Chemical Society, Chemical Communications*, 222.
Young, C.J. and Greig, H.G. (1954) *RCA Review*, **15**, 476.
Young, C.J. (1957) US Patents 2 786 439 (1957) and 2 786 441 (1957).

5 Ink jet printing
R.W. KENYON

5.1 Introduction

Ink jet printing is the only true primary non-impact printing process whereby a liquid ink is squirted through very fine nozzles and the resultant ink droplets form an image directly on the substrate (Gregory, 1991). In the main competing technologies of electrophotography (photocopying and laser printing) and thermal transfer printing, further components and steps are required to form the image, namely a photoconductor drum or belt and a transfer ribbon, respectively.

Ink jet can be conveniently divided into two main technologies, continuous and drop-on-demand. Each of these can be subdivided further as shown in Figure 5.1

The first use of ink jet in recording devices dates back to 1930 and the first successful product was developed by Elmquist in Sweden in 1951. However, it was much later that developments were made upon which today's technologies are based. In 1964 Sweet developed the continuous ink jet printing system and this was modified by Hertz in 1969 (Hertz and Heinzl, 1985). The second major ink jet technology of drop-on-demand was invented by Zoltan in 1972. This used a piezoelectric element to eject the ink. The phase-change or hot-melt system also uses the piezo technology and was first used by Data Products, formerly Exxon (Exxon, 1984) and Howtek (Howtek, 1984). Since then, Tektronix, Spectra and Brother have also been active in this technology. The Microjet technology invented by Xaar is a novel, recently patented piezoelectric-based system where the walls of the ink jet channels vibrate to eject the ink (Willis, 1991).

The thermal ink jet or bubble-jet system was discovered independently in 1979 by Canon (Lyne, 1986) and Hewlett Packard (Hammond, 1984). These two companies cross-licensed the thermal ink jet technology in 1983. It is alleged that this technology was discovered by accidentally touching a syringe needle with a hot soldering iron causing a droplet of liquid to be ejected from the syringe nozzle.

These thermal ink jet printers, which currently dominate the market-place, can employ either a permanent printhead (Canon type) or use disposable cartridges (HP type).

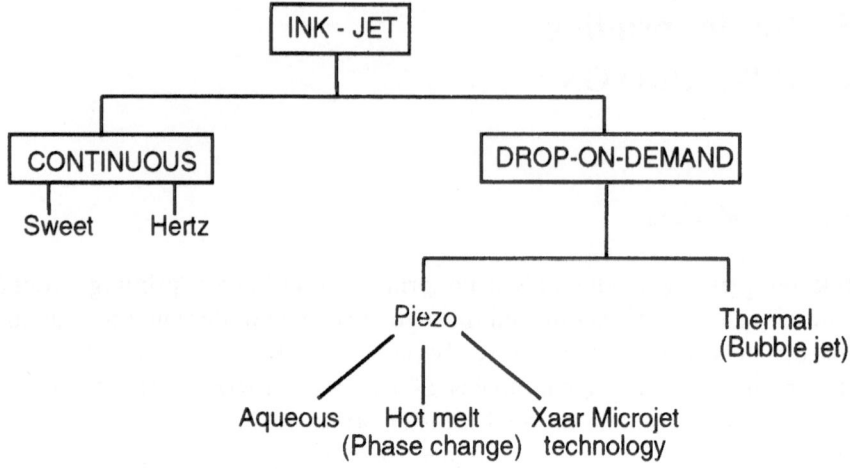

Figure 5.1 Ink jet printing technologies.

The piezoelectric printhead is more expensive to manufacture and this is also of the permanent type and is intended to last for the lifetime of the printer.

5.2 Ink jet technology

As shown in Figure 5.1, ink jet may be divided into two basic types, continuous and drop-on-demand.

5.2.1 Continuous ink jet

In continuous ink jet systems, a continuous stream of ink droplets is ejected from a nozzle. By selectively charging some of these ink droplets they can then be deflected when passing through high-voltage deflection plates. There are two possible methods of obtaining a printed image by this process. In the first method, the charged droplets are deflected on to the substrate to form the image and the uncharged droplets collected in a gutter to be returned to the ink reservoir. This is referred to as the 'raster scan method' (Figure 5.2). In the second design, the uncharged ink droplets form the image and the charged droplets are deflected to the gutter. This is the binary 'continuous ink jet system' (Figure 5.3).

Both the raster scan and binary continuous ink jet systems are based on the original Sweet technology and are especially suited to monochrome printing for industrial use. The Hertz technology, whereby a fine mist of irregular-sized ink droplets is formed, uses the binary design of the uncharged droplets forming the image (Figure 5.4). It is more suited to

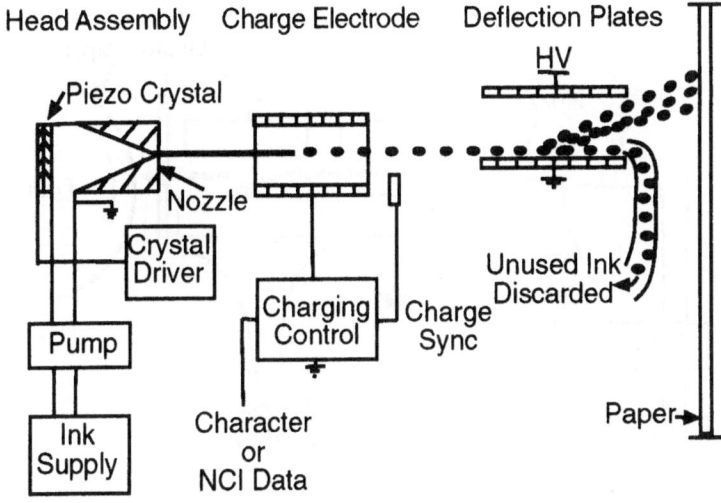

Figure 5.2 Raster scan continuous ink jet.

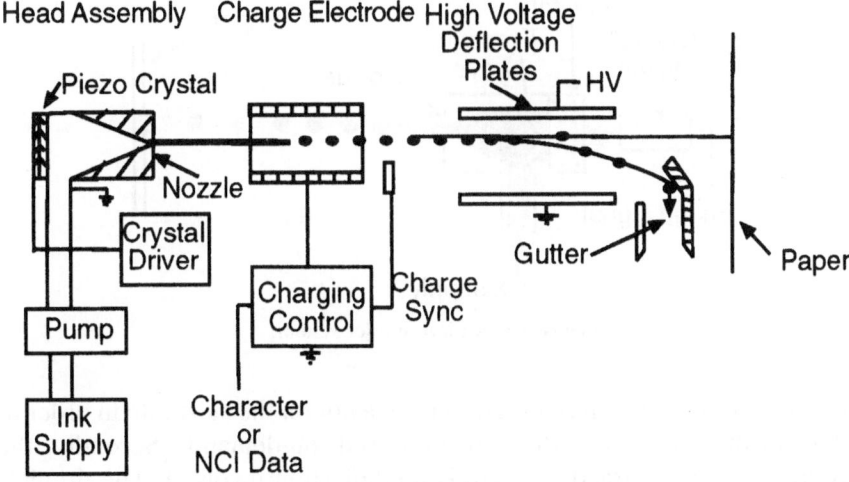

Figure 5.3 Binary continuous ink jet.

colour printing than the previous two methods and is employed by Iris in their colour ink jet printer.

5.2.2 *Drop-on-demand ink jet*

Drop-on-demand ink jet systems differ from continuous ink jet in two major respects. Firstly, every ink droplet produced is used to form the image. No droplet is wasted or returned to the ink reservoir. Ink droplets

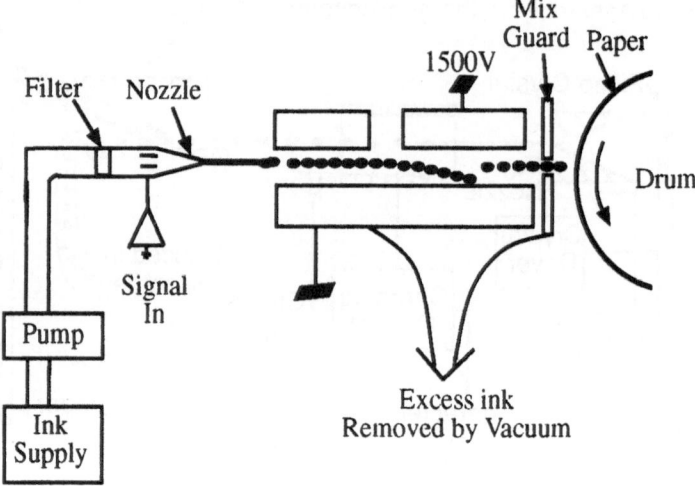

Figure 5.4 Hertz continuous ink jet technology.

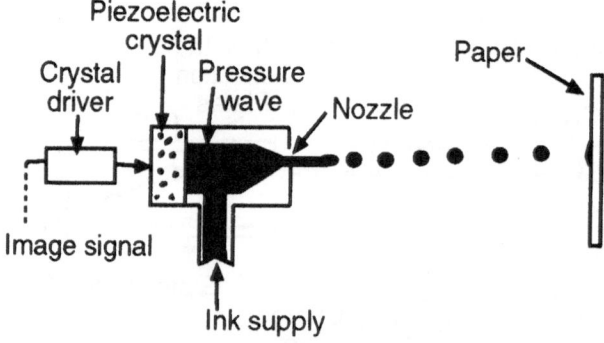

Figure 5.5 A piezo ink jet system.

are only produced when required to form a dot on the substrate in order to build up the image, i.e. they are produced 'on-demand'. Secondly, the droplets are not charged so there is no deflection involved. The droplets are fired in a straight line to the desired point on the substrate. The printhead should therefore be placed as close as possible to the substrate so that the distance travelled by the droplets is minimal. This allows more accurate positioning of the dots leading to high-quality prints. This makes drop-on-demand printers suitable for office printing. Continuous ink jet printers are more suitable to industrial use since the droplets can travel further distances.

Drop-on-demand ink jet printers can be subdivided into two main types, piezo and thermal (bubble jet). Piezo ink jet is one of the simplest forms of drop-on-demand printing where ink droplets are ejected by the action of an oscillating piezo crystal (Figure 5.5).

The first generation of piezo ink jet printers appeared in the marketplace in the late 1970s. Silonics, Siemens and Toray were amongst the first in this area. In the early 1980s, numerous companies introduced printers of this type including Canon, Data Products, Exxon, Ricoh, Sharp and Tektronix. More recently, Seiko Epson have launched a number of very successful printers (Stylus range) based on this technology.

A recent variation of the piezo system is the Microjet technology invented by Xaar. In this adaptation, the walls of the ink jet channels vibrate to eject the ink (Figure 5.6). Claimed advantages of the Microjet technology are ease of manufacture, constant drop size and velocity, high resolution and the potential to produce page-wide arrays.

Phase change (or hot melt) ink jet also uses the piezo technology. In this system, the ink, which is solid at ambient temperature, requires to be heated and kept molten in the printhead. It can then be ejected through the nozzles and upon hitting the substrate it cools and resolidifies (Figure 5.7).

The main advantages of phase-change ink jet printing are excellent waterfastness and good substrate independence. However, phase-change ink jet printers are somewhat more expensive to manufacture when compared to other impulse ink jet technologies and running costs are higher due to the energy required to maintain the ink at elevated temperatures. Poor adhesion and height of the print on the paper (Braille effect) were other disadvantages when this technology was first introduced. However, these have been partly resolved by ink formulation, and subjecting the print to a cold-fusion technique. Nonetheless, the prints still

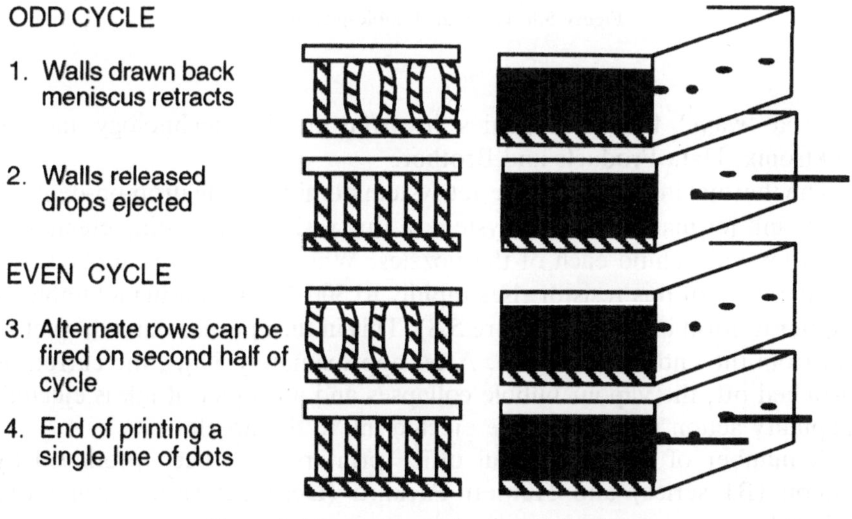

ODD CYCLE

1. Walls drawn back meniscus retracts

2. Walls released drops ejected

EVEN CYCLE

3. Alternate rows can be fired on second half of cycle

4. End of printing a single line of dots

Figure 5.6 Xaar piezo technology (Microjet).

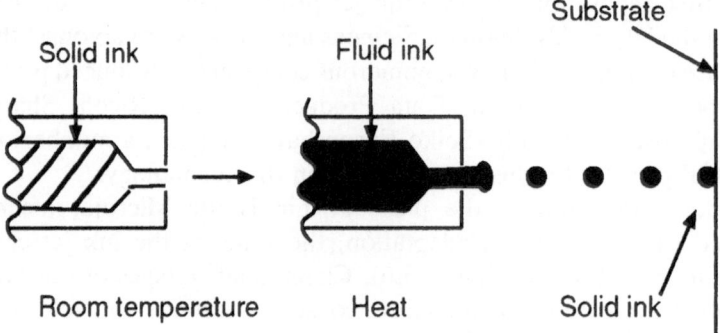

Figure 5.7 Phase-change ink jet.

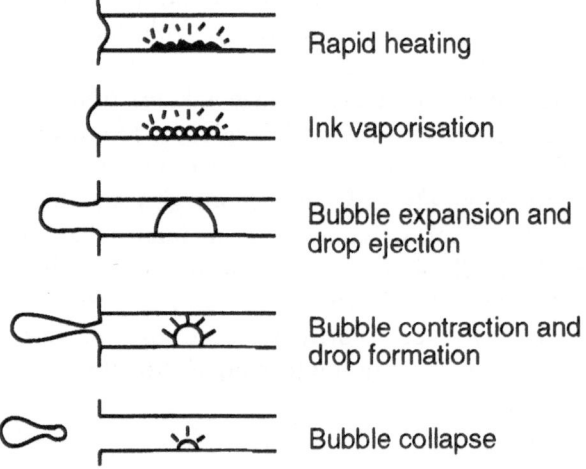

Figure 5.8 Thermal (bubble-jet) ink jet.

have a 'waxy' feel. Companies involved in this technology include Tektronix, Data Products and Brother.

The thermal ink jet or bubble-jet system dominates the drop-on-demand office ink jet market. In this system, a tiny resistor or heating element is incorporated behind each of the nozzles. When a current is applied, the temperature of this resistor rises rapidly to 300–350°C causing a bubble of vapour to form in the ink (Figure 5.8). This in turn pushes out a droplet of ink from the end of the nozzle. After approximately 50 μs, the current is switched off, the vapour bubble collapses and a droplet of ink is ejected. Capillary action replenishes the ejected ink in the nozzle.

A number of very successful office printers have been produced by Canon (BJ series) and Hewlett Packard (Desk Jet series) using this principle.

5.3 Ink jet inks

There are basically three types of ink jet inks: aqueous, solvent and phase change. The most common inks for drop-on-demand office printers are aqueous-based whereas solvent-based inks are used mainly with continuous printers for industrial use. However, the Xaar Microjet technology uses a high-boiling point solvent with low vapour pressure as the ink vehicle. The phase-change inks are based on low melting point waxes and resins. Before considering these inks individually, some general features required of all ink jet inks are discussed.

5.3.1 General features

The most important factors of all ink jet inks are high reliability, the ability to produce good print quality and compatibility with the printer. Table 5.1 shows the important properties that influence droplet formation, the printing properties required of an ink to produce good quality prints and the more general properties required of an ink including compatibility with the printer.

5.3.2 Aqueous inks

In addition to water, aqueous-based inks normally contain water-miscible co-solvents. These are usually glycol types, e.g. diethylene glycol. Other solvents such as pyrrolidones are also common. These co-solvents serve various functions. For example, they act as humectants, i.e. they help to minimise the evaporation of water and thus prevent crystallisation or crusting of the dye at the ink jet nozzles. Any crusting would block the

Table 5.1 Important properties of ink jet inks

Physical properties	Printing properties	Functional properties
Viscosity	Good optical density	Machine compatibility
Specific gravity	No feathering	Stable drop formation
Surface tension	Uniform spreading	No particle contamination
Dielectric properties	Fast set time	No nozzle crusting
pH	Good waterfastness	Low corrosion
Colour	Good fading resistance	No biological growth
	Good soak resistance	Long shelf-life
	Good rub resistance	No long-term health
	Good smear resistance	problems
	Gloss	No chemical hazards
	Able to print on paper variations	

nozzles and prevent the system from operating correctly. Co-solvents can also aid in controlling the viscosity and surface tension of the inks. In certain cases, they can also improve the solubility of the dye, producing more stable inks.

Water is a good medium for the growth of microorganisms such as bacteria and fungi and if left unchecked this could lead to blocking of the nozzles. A biocide effective against both bacteria and fungi is therefore normally added to the ink. Certain co-solvents such as alcohols or glycols can also serve as the biocide.

Water also has the disadvantage of being a good corrosion promoter and this can be accentuated by the presence of certain electrolytes, e.g. chloride ions, in the dye. Great care must be taken to select both a pure dye and a pH that is compatible with the materials used in the ink and in the printer.

Insoluble material in the ink can also present problems and inks must be filtered before use. The main source of such contamination is the dye itself and only highly purified dyes should be used. Chelating agents are sometimes added to solubilise certain metal salts, e.g. iron compounds.

The majority of colorants used in aqueous ink jet inks are water-soluble dyes to produce true solutions. However, a number of patents have now appeared claiming the use of stable pigment dispersions in aqueous ink jet inks. These dispersions also include a resin in the formation in order to fix the pigment on to the substrate and a product of this type using carbon black as the pigment has now been commercialised by Hewlett Packard in their 1200C printer.

5.3.3 Solvent inks

Solvent-based inks are mainly encountered in the industrial segment of ink jet printing. These are used when fast drying times are required and/or when printing on to hydrophobic substrates such as metals, plastics or glass. Traditionally, ketonic solvents, e.g. methyl ethyl ketone (MEK), have been used in this area. However, a major disadvantage of such solvents is flammability and because of this, alternative alcoholic solvents such as ethanol and n-propanol are now being used. These are less flammable than ketonic solvents and faster drying than water. Solvents also pose more of a toxicity problem than water and some solvent inks are already being banned in certain parts of the USA.

The predominant colour in industrial marking is black and for solvent-based inks, solvent-soluble dyes are required. However, in the Xaar Microjet system a pigment dispersion (carbon black) in a high boiling solvent is used.

5.3.4 Hot-melt inks

Hot-melt inks are similar to wax crayons, being solid at room temperature and molten when heated. The preferred melting points are in the range 60–125°C. The vehicles used appear to be either long-chain (C_{18}–C_{24}) fatty carboxylic acids or alcohols as claimed by Exxon, or sulphonamides as claimed by Tektronix. The colorants used are solvent dyes that are soluble in the particular ink vehicle, or pigment dispersions.

5.4 Ink jet colorants

The majority of office ink jet printers, both thermal and piezo, use aqueous inks. Consequently, the colorant, apart from one exception, is a water-soluble dye. As is the case with other printing technologies, a black and three subtractive primary colours of yellow, magenta and cyan are required.

The early ink jet printers had a reputation of unreliability, this being almost certainly due to the dyestuffs used. These first-generation dyes were 'off-the-shelf', readily commercially available products, usually direct dyes, food dyes and acid dyes. The direct and acid dyes as manufactured for textile use are not pure and, in addition, other chemicals are added to give the final formulated product. Many of the dyes also had inadequate solubility for ink jet purposes. The result of this was failure of some or all of the nozzles to eject ink due to clogging by particulate matter in the ink. This insoluble material could be crystallised dye or an impurity.

5.4.1 Black dyes (first-generation)

The unreliability problem was solved by using a very water-soluble black dye, CI Food Black 2 (1). Food Black 2 is a simple disazo dye containing four sulphonic acid (SO_3H) groups, the best substituent for imparting water solubility to a molecule. In addition, being a food dye, it is toxicologically safe. After purification to ink jet dye quality, Food Black 2 solves the reliability problem and gives attractive neutral black shades.

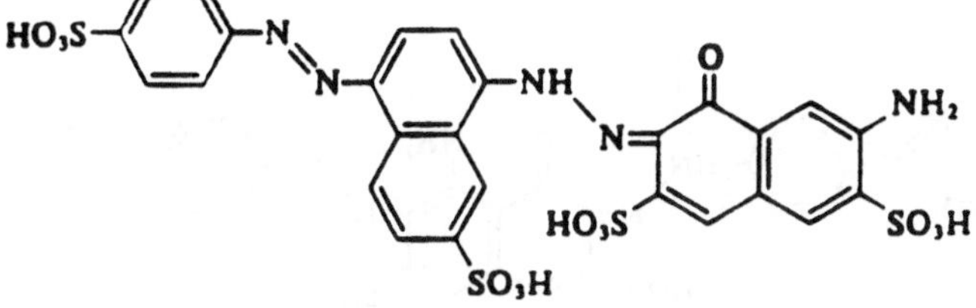

(1)

However, Food Black 2 does have one major problem, that of poor waterfastness of the prints, especially when printing on plain papers. The reason for this is obvious. The dye forming the image on the paper is still very water soluble. If the print comes into contact with water then the dye will redissolve, leading to severe smudging. The most common cause of smudging is from handling with moist fingers, smearing with a highlighter pen or spillage of water, tea, coffee, etc.

Other black dyes that have been used in ink jet printing include the trisazo dyes CI Direct Black 154 (2), CI Direct Black 171 (3) and CI Direct Black 168 (4). The tetrakis azo dye CI Direct Black 19 (5) may also be used to give better waterfastness but the aqueous solubility is low.

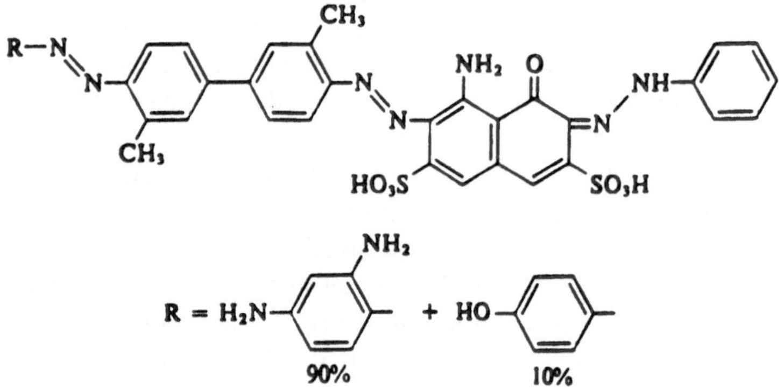

(2)

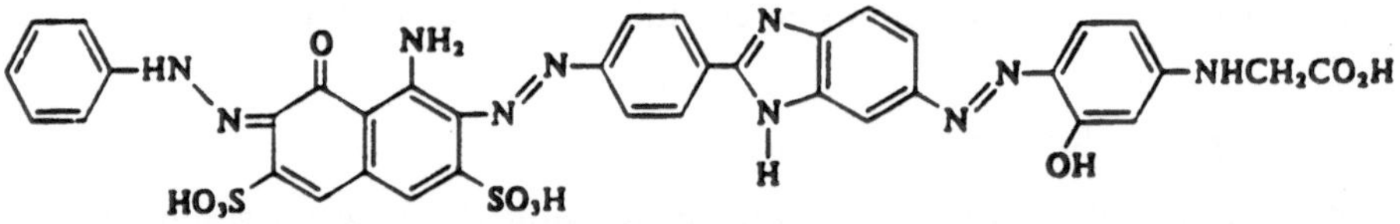

(3)

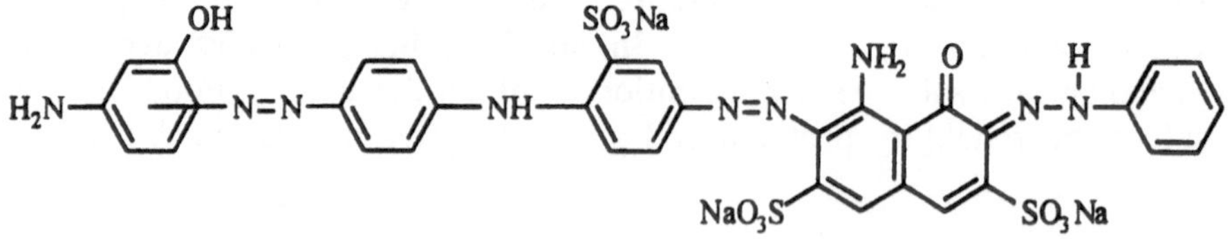

(4)

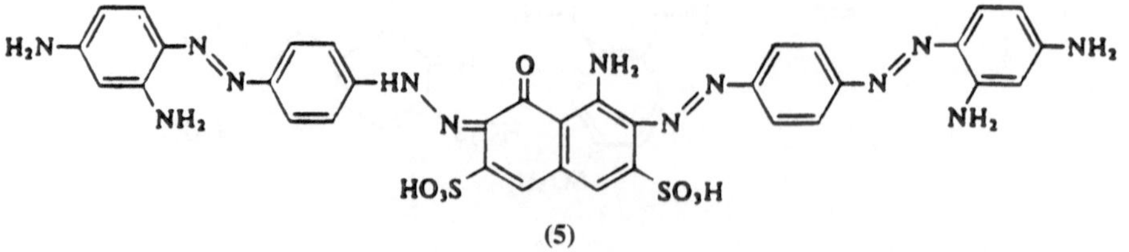

(5)

Being more complex structures, all the tris- and tetrakis-azo dyes exhibit thermal instability leading to problems of kogation (charring of insoluble deposits on the heating elements). They are thus more suited to piezo ink jet printers than to thermal printers.

5.4.2 Colour dyes (first generation)

As in the black-shade area, the first-generation colour ink jet dyes were also 'off-the-shelf', commercially available products purified to ink jet standards. There are two established colour trichromats: one based on silver halide photography and one based on process colours for magazine and newspaper printing. The photographic trichromat uses a reddish-yellow dye since one of the major uses of photography is taking pictures of people and skin tones have to be faithfully reproduced. Consequently, a bluish magenta and a greenish cyan are required. In process colour printing, people are not the main subjects and this trichromat employs a greenish yellow, a redder magenta and a bluer cyan. No standard trichromat has yet emerged in ink jet printing but the colours, especially the yellow, are more representative of the process colours rather than the photographic colours.

5.4.2.1 *Yellows.* The three main first-generation ink jet yellow dyes used are CI Direct Yellow 86 (**6**), CI Direct Yellow 132 (**7**) and CI Acid Yellow 23 (**8**).

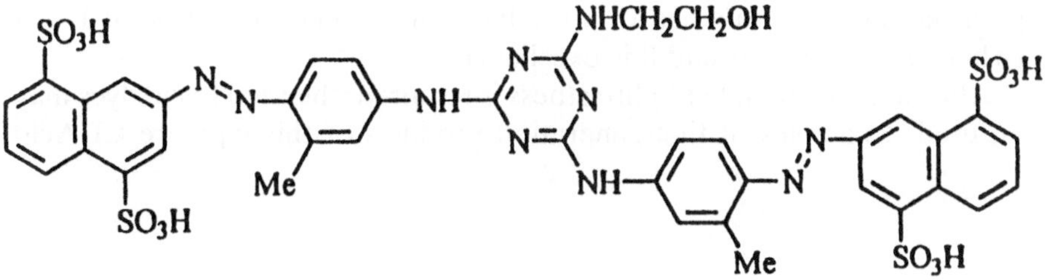

(6)

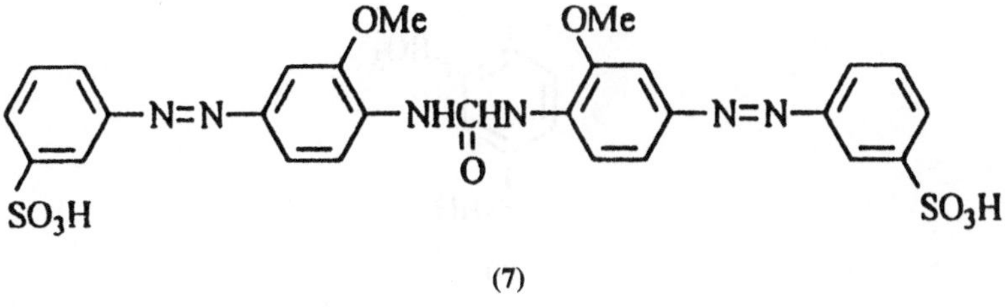

(7)

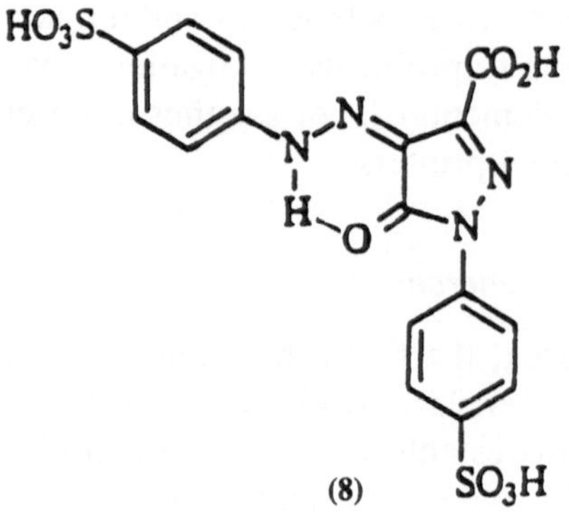

(8)   SO$_3$H

CI Direct Yellow 86 is a reddish-yellow dye typical of a photographic yellow. It has good all-round properties but is generally considered to be slightly too red in shade. CI Direct Yellow 132 is slightly greener and gives very attractive mid-yellow shades. CI Acid Yellow 23, also known as tartrazine, is based on a pyrazolone coupling component and exists totally in the hydrazone form shown. Acid Yellow 23 has inferior lightfastness than both Direct Yellow 86 and Direct Yellow 132. Being a monoazo dye it also exhibits inferior waterfastness. It is the most greenish yellow of the three dyes.

5.4.2.2 *Magentas.* Very bright magenta ink jet prints can be obtained using xanthene dyes such as CI Acid Red 52 (9). Dyes of this type give brilliant, usually bluish magenta shades but unfortunately they exhibit very poor lightfastness, rated as 1 on the logarithmic blue wool scale of 1 to 8 (where 1 is very poor and 8 is excellent).

Alternatively, if higher lightfastness is required, the azo H-acid dyes may be used. Examples of first-generation products of this type are CI Acid

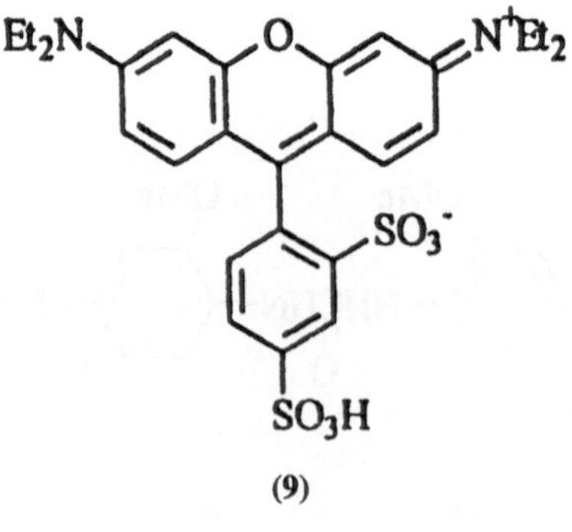

(9)

Red 249 (**10**), the dye (**11**) and dyes of structure (**12**). All these dyes exist predominantly in the hydrazone form. Even so, the lightfastness of most of these products is rated at approximately 3 (some four times better than Acid Red 52).

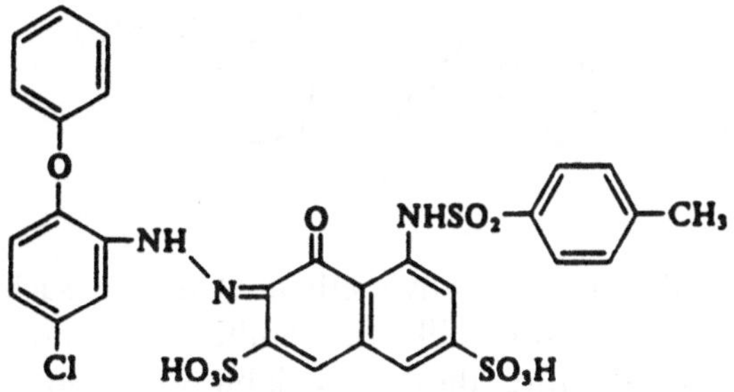

(10)

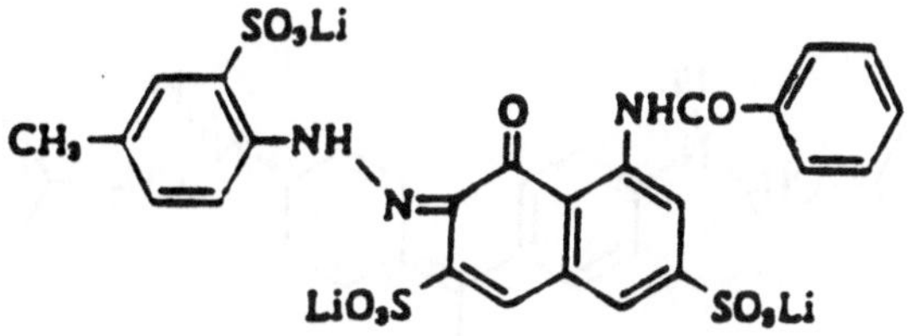

(11)

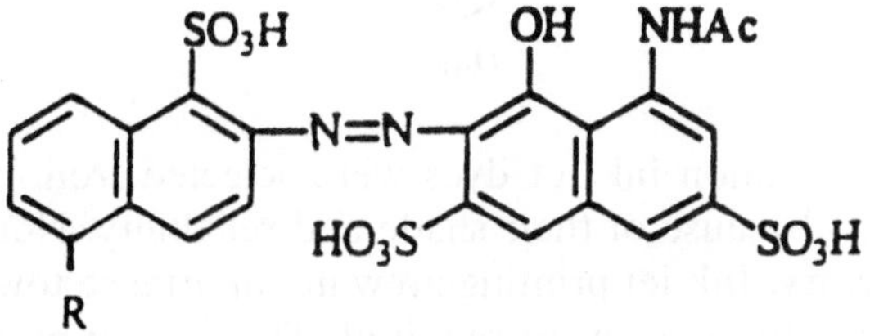

(12)

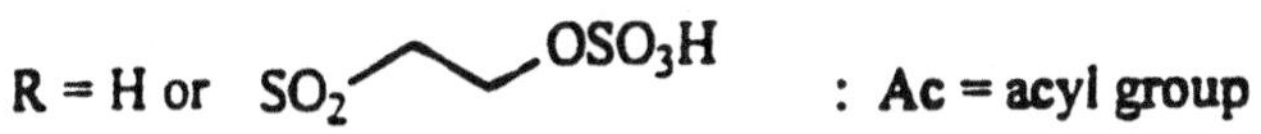

$$R = H \text{ or } SO_2\text{—}CH_2CH_2\text{—}OSO_3H \qquad : Ac = acyl\ group$$

5.4.2.3 Cyans. There are two main classes of dyestuff used for the first-generation cyan products: phthalocyanines and triphenylmethanes. The most important of these are dyes based on copper phthalocyanine. These exhibit bright cyan shades having very high durability to light, heat and chemicals and are economical to manufacture. An example of this type is CI Direct Blue 199 (**13**).

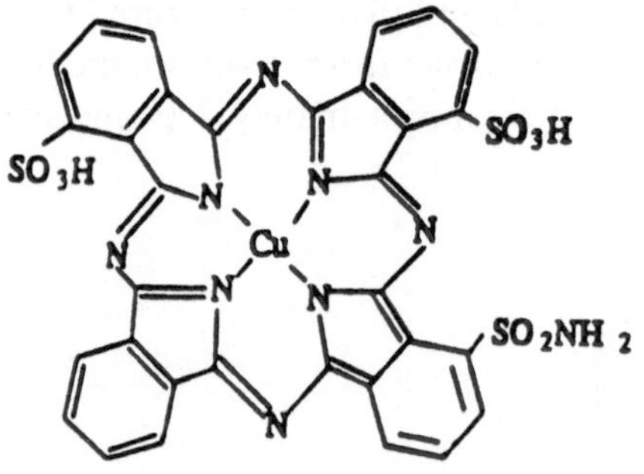

(13)

Cyan dyes based on the triphenylmethane structure, such as Acid Blue 9 (14), are very bright. However, like the xanthene magenta dyes which they closely resemble in structure, they also exhibit poor lightfastness (approximately 1–2 on the blue wool scale). By comparison, copper phthalocyanine dyes exhibit a lightfastness of at least 4.

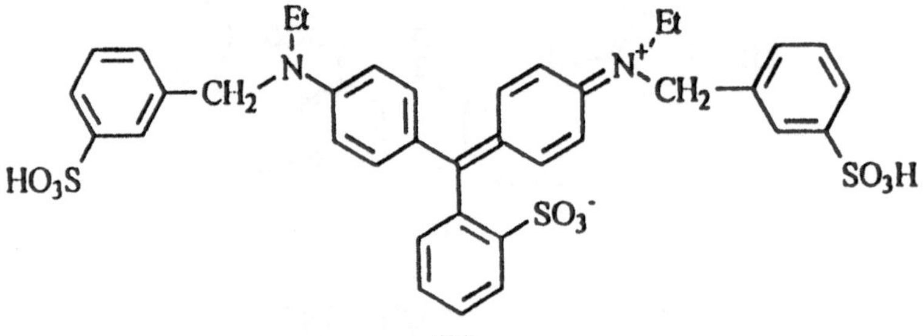

(14)

All the first-generation ink jet dyes were selected from commercially available products because of their shade and reliability. None, however, was fully satisfactory. Ink jet printing grew in importance towards the end of the 1980s and better dyes were required. The desired properties could only be achieved by synthesising completely new molecules especially for ink jet. These second-generation dyes are now discussed.

5.5 Second-generation colorants

A number of companies throughout the world have been involved in research into new colorants for use in ink jet printing. Japanese companies include Mitsubishi Chemical, Canon, Orient and Ricoh; US companies include Xerox, Lexmark, Tektronix and DuPont whilst in Europe, Zeneca, BASF and Hoechst have been involved. This has led to a number of new products being introduced into ink jet printers.

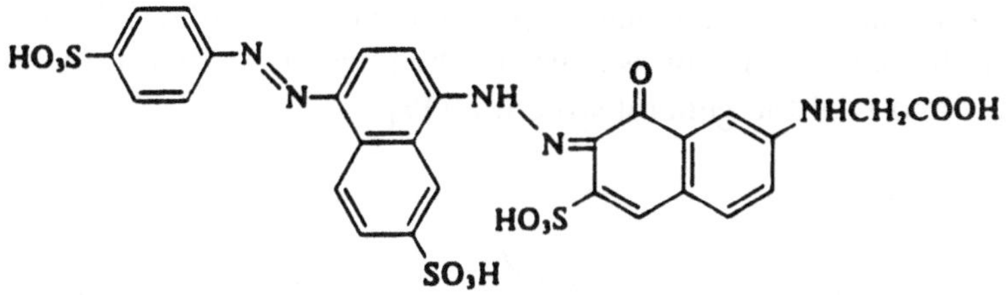

(15)

Research by Mitsubishi and Canon led to a modified Food Black 2 structure (15) although waterfastness was not significantly improved.

A step-change in the waterfastness of ink jet dyes arose from a collaboration between ICI (now Zeneca) and Hewlett Packard in 1987. The approach used by Zeneca was that of differential solubility whereby a dye was used that had high solubility in the aqueous ink media but low solubility when on the paper. Aqueous ink jet inks are normally slightly alkaline (in the pH range 7.5–10) whereas most plain papers are slightly acidic (pH 4.0–6.5). All of the first-generation dyes used in ink jet printing contain an abundance of strongly acidic sulphonic acid groups as the water-solubilising substituents and these dyes are often still soluble even under acidic conditions. By selectively replacing some of the sulphonic acid groups by less acidic substituents, e.g. carboxylic acid, then dyes can be produced that have high solubility in an alkaline ink (where the carboxy group is ionised and therefore confers extra water-solubilising properties) but which have relatively low solubility on the acidic paper. The carboxylic acid group would then be present in its less water-soluble non-ionised form. Dyes of this type are now used in several of the Hewlett Packard printers: they are based on the Food Black 2 structure and have the general formula (16).

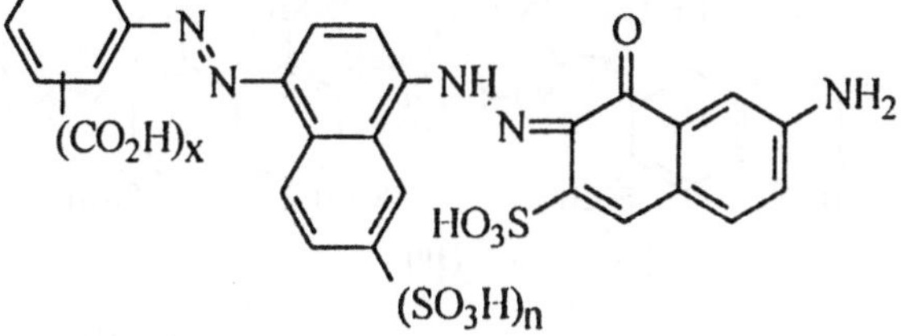

$n = 0$ or 1
$x = 1$ or 2

(16)

More recently, another black dye employing this differential solubility concept has appeared. This is marketed by Lexmark in their EXECJET 11 printers and is of the general structure (17).

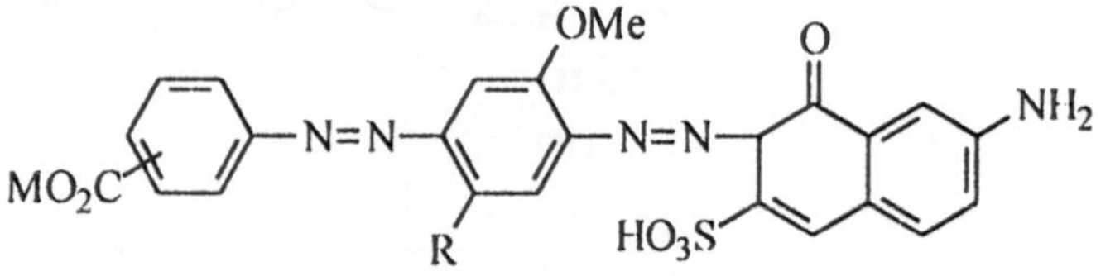

(17)

The only aqueous-based office ink jet printer that does not use a water-soluble dye is Hewlett Packard's 1200C. This uses a pigment-based ink, namely carbon black. This arose from a collaboration between Hewlett Packard and DuPont, the ink also containing a resin in order to 'glue' the pigment on to the paper. These heterogeneous inks are more complex where very stable dispersions are required and the resins used need to be free of kogation problems when used in thermal ink jet printers and must not cause any clogging of the nozzles.

The only novel colour dyes to be launched are again based on the differential solubility approach from research carried out by Zeneca (Kenyon, 1994). This high waterfast trichromat of yellow (18), magenta (19), and cyan (20) dyes appeared in certain printers in late 1994.

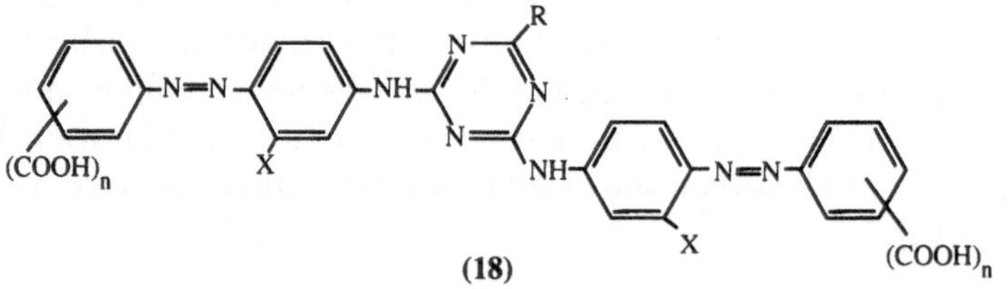

(18)

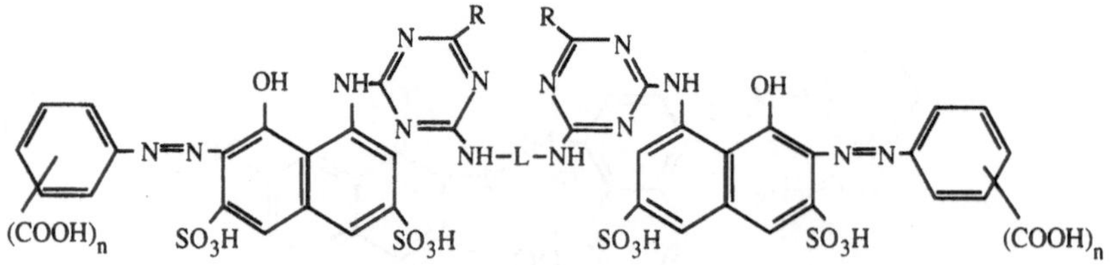

(19)

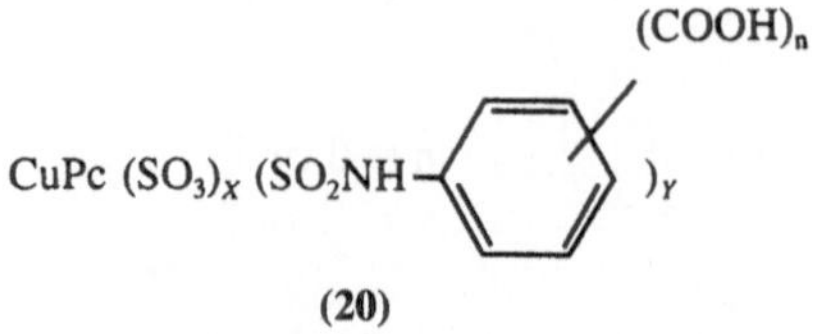

(20)

5.6 Phase-change colorants

The colorants used in phase change inks may be either dyes or pigments. The dyes chosen must be soluble in the ink vehicle, a wax or resin. Thus, solvent dyes are normally employed, such as CI Solvent Black 35 (21). Typical pigments that may be used are carbon black, CI Pigment Yellow 12 (22), CI Pigment Red 57:1 (23) and CI Pigment Blue 15:3 (24).

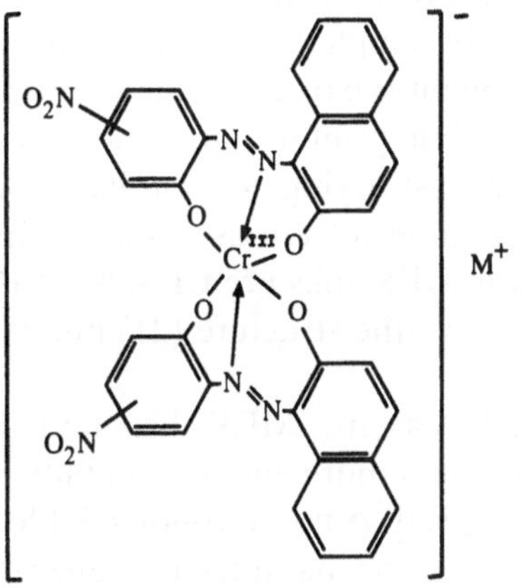

(21)

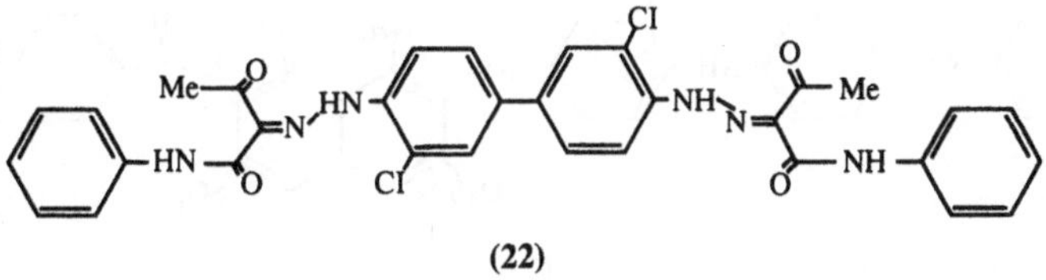

(22)

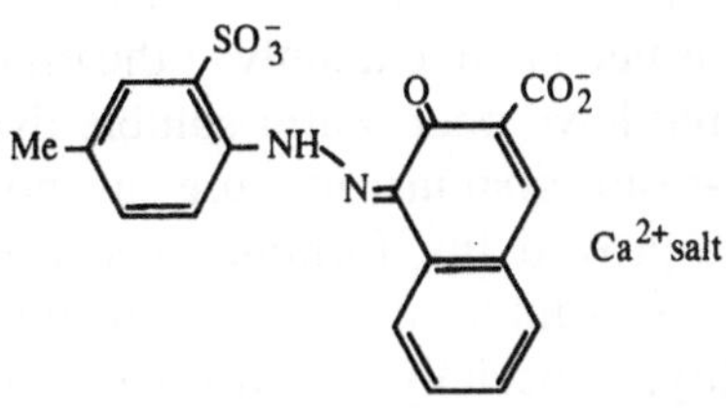

(23)

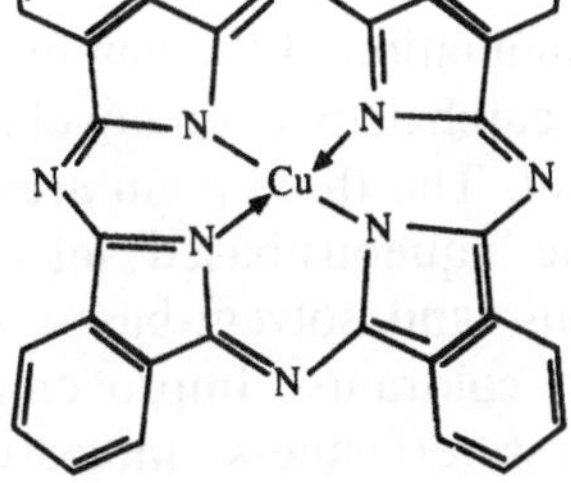

(24)

5.7 Colorants for industrial ink jet printers

Industrial ink jet printers are normally of the continuous type and are situated adjacent to conveyor belts. It is a very good method of printing an identification mark on to packages and products as they quickly pass by along these conveyor belts. These markings can be, for example, batch numbers, bar codes, sell-by dates, etc. These industrial printers can print on to a variety of materials such as paper or cardboard, plastic or metal. By far the most important colour in industrial printing is black.

For printing on to porous substrates such as paper or cardboard, water-based inks are normally employed. Typical dyes are MOBAY Black SP liquid, a three-component mixture of yellow, red and navy blue dyes, and CI Direct Black 168. For printing on to non-porous substrates, such as metal or plastics, a fast-drying solvent-based ink is required. Until recently, the most common solvent used was methyl ethyl ketone (MEK). Typical black dyes for MEK inks are CI Solvent Black 29 and CI Solvent Black 35. Both dyes have the structure (21) but contain different ratios of the two isomers.

For environmental reasons, MEK has now become less favoured, especially in the USA, and more environmentally friendly solvents such as ethanol are now being used. A new alcohol-soluble black dye (25) based on Direct Black 168 is now being marketed by Zeneca for use in these alcohol-based inks.

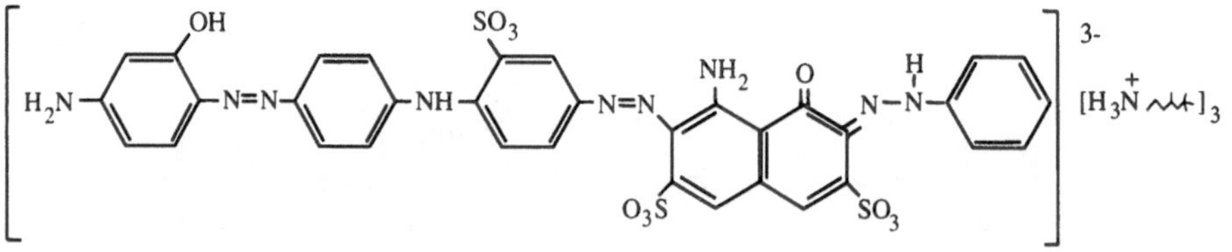

(25)

5.8 Ink jet ink/colorant research

By examination of the patents published on ink jet over the last ten years, the research activity on novel colorants has been aimed mainly at the office segment. The three main areas of research have been water-soluble dyes for the aqueous-based inks, pigment–resin systems for use in both aqueous- and solvent-based inks and improved ink formulations using known colorants. Improvements in water-soluble dyes have included better waterfastness, increased durability, especially lightfastness, and colour dyes with superior chroma. Pigment–resin systems have claimed advantages in waterfastness, durability and the ability to print on a wide variety of substrates. However, only one printer, Hewlett Packard's 1200C, has so far emerged using a pigment–resin ink, this being based on carbon black. The colour inks in this printer, yellow, magenta and cyan, still use conventional water-soluble dyes. With colour ink jet printing becoming more and more important a colour trichromat based on pigment inks is still awaited.

The research into ink formulations has been mainly aimed at improvement in print quality, for example less feathering and superior black–colour and colour–colour bleed. In addition, the use of various additives and new humectants have also been claimed in order to improve the solubility of the dye to prevent nozzle clogging or crusting.

Ink jet is currently at a crossroads as regards both colorant and vehicle. Whether a dye or pigment is chosen for a particular application has been recently reviewed (Gregory, 1994; Work, 1994). This will depend on how properties such as colour gamut, transparency, colouring power, choice/ flexibility and ease of use (dye attributes) are ranked against excellent lightfastness, insolubility and crystallinity (pigment attributes). In general, dyes not only equal the properties of pigments but also offer many other additional benefits.

5.9 Advantages/limitations of ink jet

In the past, hard copies in offices were always produced using traditional impact printers such as the typewriter, the daisywheel and more recently the dot-matrix printer. However, in general, these tended to be rather slow and quite noisy machines.

With the arrival of computer terminals and word processors in the 1980s, offices underwent a revolution and electronically controlled non-impact printers such as the ink jet and laser printer are now commonplace in most offices. By comparison, these are very fast and almost silent in operation, able to produce both text and graphics and also showing full-colour printing capabilities.

The main advantage of ink jet over other non-impact printing techniques is its low cost and simplicity. As mentioned previously, it is the only true primary non-impact printing process whereby a liquid ink is squirted directly on to the substrate to form an image. Laser printers require a photoconductor belt or drum (see Chapter 4) and transfer ribbons are necessary in thermal transfer printing (see Chapter 6).

Ink jet printers are now rapidly displacing the dot-matrix printers because of their superior technical performance at reasonable cost. Penetration of the lower-end laser printer market is also now occurring because of the cost advantage. The price of a monochrome laser printer is approximately twice that of an ink jet printer and for colour printers this price gap widens considerably. The main advantage of laser printers, however, is print quality, the early ink jet printers exhibiting very poor waterfastness on plain papers. Recent advances involving novel dye synthesis have now essentially overcome the problem and black and colour dyes having very high waterfastness are now available. Advances in formulation have also improved print quality, especially colour–colour bleed, and full-colour prints suitable, for example, for presentations can now be obtained from recently introduced colour ink jet printers.

An ink jet printer is, however, currently much slower than a laser printer. Being aqueous based, the dry time of an ink jet print, especially full-colour prints, can be quite long. In addition, cockle and curl problems can also occur. Further advances in ink jet inks are required in order to overcome these problems.

Thermal transfer printing is also much more expensive than ink jet printing and requires special receiver sheets and dye sheets. However, very high print optical densities can be obtained with this technology and continuous tones (grey scale) can be obtained without loss of resolution leading to high print quality (see Chapter 6). Thermal transfer prints are thus more comparable to conventional silver halide photographic prints (see Chapter 3). To compete in these markets, ink jet requires improvements in the implementation of dot-size modulation for the simulation of continuous tone image production. This would allow a much improved perception of print and image quality.

5.10 Uses of ink jet printers

5.10.1 Office segment

Over the last few years there has been a growing demand for page ink jet printers with colour capabilities to be used in conjunction with business and home-based personal computers. Due to the enormous amount of text printed in this segment, the main shade required is black. However, full-colour printing is now becoming very important and the colour printers now being introduced into the marketplace will shortly outsell monochrome printers worldwide.

In this office area, the drop-on-demand printers predominate. With higher resolutions now achievable (600 × 300 dots per inch becoming the entry point for ink jet), near letter-quality text and high-quality graphics can now be produced on the majority of office papers. The office ink jet printers are thus ideal for the preparation of short business reports or communications, possibly containing some full-colour printing. However, for long-run prints, ink jet is currently uneconomical compared to conventional printing technologies such as offset lithography and flexo-graphy (see Chapter 2).

Another major use of the office ink jet printer is the printing of transparencies for overhead projectors. Bright colours are achievable and give excellent results for use in presentations. The overhead transparencies are designed specifically for ink jet printing, and definition and drying time have been improved by incorporating a 10–30 μm hydrophilic layer at the surface of the polyester film.

One of the major disadvantages of facsimile (fax) printers in the past has been the necessity to use special thermal papers. Ink jet facsimile machines are now available producing high-quality prints on plain paper. This is an area which could show considerable growth in the near future.

For the wide-format ink jet prints used for graphic arts, signage and display purposes, the print is often protected against ultraviolet radiation, moisture, fungus growth, etc. which would cause the images to fade or discolour and the substrate to deteriorate. For this purpose, laminated films are used both for indoor and outdoor environments and these can contain ultraviolet absorbers.

5.10.2 Industrial segment

Although the industrial ink jet printers require dyes of a higher purity than standard commercial textile dyes, they need not be to the exacting standard of the office printers. The resolution is also much lower than in the hi-tech office segment. The inks used can be aqueous-based for use on porous substrates such as paper or cardboard, or solvent-based for substrates such as plastic or metal. If opacity is required, for example when printing on to black substrates, then pigment inks are necessary. There are a number of companies involved in this industrial segment including Diagraph, Domino, Image and Videojet.

Industrial ink jet printing has been very successful in meeting the needs in the 'best before' and 'sell by' coding of packages and products, especially on delicate or non-flat surfaces. In these applications, ink jet demonstrates considerable advantages over conventional techniques such as stamping, stencilling etc. Because ink jet devices are non-contact, they do not impede the product flow in the same way as other non-impact devices, e.g. electrophotographic printing, where the need for toner transfer and subsequent fusing limit the speeds at which imaging can take place.

Ink jet systems have been used in the mailing industry for some years and have also been used for 'lucky numbering' of newspapers and periodicals. Other uses include the numbering and bar coding of labels, tickets, tags, etc.

A potential threat to industrial ink jet printing is direct laser imaging where identification marks can be introduced on to, for example, metal cans directly, using a laser. This type of technology could replace the solvent-based ink jet area, now considered by a number of countries to be an environmental hazard.

5.10.3 Colour filters for LCDs (Liquid crystal displays)

There are examples in the more recent patent literature which describe the use of ink jet printing in the production of colour filters for LCDs. However, no manufacturing procedure is currently used on a commercial scale.

In all cases, appropriately coloured inks are deposited on to glass which has been coated with a special transparent 'receiver' layer. Materials which find application in this area include resins such as acrylates, polymers such as polyesters, polyamides and polyvinyl alcohol and others such as casein and gelatin. Many ink-receiver layers contain additives to improve the robustness of the derived colour filter, such as cross-linking agents (melamine-formaldehyde adducts) or cationic resins.

A major deficiency of many commercial colour-filter manufacturing processes is the lengthy production time due mainly to the large number of different stages involved. Several of the ink jet processes in the literature describe the use of a mask or grid to control ink deposition and hence facilitate the patterning of the variously coloured pixels. The future may see enhancements in these methods, perhaps exploiting the inherent accuracy of the ink jet technique in conjunction with specially designed inks which can be deposited directly on to the glass itself, thus reducing the length and complexity of manufacture.

5.10.4 Ink jet printing of textiles

One of the major bottlenecks in textile printing is the pre-print and sampling stage. Before a customer can receive a design on a textile substrate, the printer has firstly to produce engraved screens for each colour of the design. The fabric can then be printed, steamed (for fixation) and finally washed. This is a lengthy and expensive process. For a number of years CAD (computer-aided design) systems have been available and hard copies using ink jet printers produced. The hard-copy output was only on paper and used conventional ink jet dyes. A textile hard copy would obviously be more useful in design interpretation and recent developments have been aimed at producing hard-copy output from CAD systems directly on to textile substrates using ink formulations based on textile dyes (Provost, 1994).

The first commercially available system using textile dyes and ink jet printing technology was launched by Stork at the ITMA 1991 in Hanover. This incorporated highly purified reactive dyes into the ink formulation;

the ink jet printed sample could then be processed in a similar manner (fixation and wash-off) as a conventional screen or roller print. Samples of fabric can thus be obtained quickly in a choice of colours without having to make films, screens or print paste so customers can make a 'go/no-go' decision for a particular design or colour combination. This speeds up the pre-print phase and more informed decisions can be taken.

Further developments in the textile ink jet printing area are also taking place in Japan with textile companies cooperating with computer-electronic manufacturers. One such venture is between Canon (computer side), Kanebo (textile side) and Toshin Kogyo (textile machinery side).

For an ink jet print to be comparable with a textile print produced by conventional means, the ink formulation must make use of the same dye chemistry. Stork of Holland and Zeneca Specialties have worked closely to develop highly purified versions of certain reactive dyes and incorporated these into formulations that are compatible with the requirements of the Stork 'TruColor' jet printer. These dyes are of the monochloro-s-triazinyl type and these react under hot alkaline conditions with cellulose to produce a covalent dye–fibre bond which gives prints of excellent fastness (equation 5.1).

$$dye-NH{-}\underset{\substack{\\ \text{NHR}}}{\overset{\substack{\text{Cl}}}{\bigtriangleup}} + HO-cellulose \xrightarrow{\text{alkali}} dye-NH{-}\underset{\substack{\\ \text{NHR}}}{\overset{\substack{\text{O}-\text{cellulose}}}{\bigtriangleup}} + NaCl$$

$$(5.1)$$

The Stork 'Trucolor' jet printer is a development based on the Hertz continuous ink jet technology. Because of the stringent purity requirements required by continuous ink jet printers and the very specific conductivity specifications, none of the conventional printing chemicals (alkali, urea, sodium alginate thickener, etc.) can be incorporated into the ink formulation. However, the dyes used are conventional reactive dyes purified to ink jet standards.

Ink jet printing of hydrophobic fabrics such as polyester is also being investigated. Although reactive dyes cannot be fixed directly on to polyester, a pretreatment technology is being developed to give a realistic colour appearance on these substrates. The approach being evaluated is to introduce a hydrophilic receiving layer on the polyester to which the reactive dye would fix.

5.11   Toxicology and the environment

New legislations throughout the world are becoming more demanding and assertive about the environmental performance of the chemical industry and ink jet printing must be considered part of this industry (Elliot, 1994b). This will impact on every part of the business, from raw-material selection to manufacture of ink jet printing products, their use or application and finally to their disposal. As non-impact printing technologies penetrate the conventional inks market, they will become subject to control from an increasing number of existing and future Environment and Safety Directives.

An important part of the assessment of any new ink for safety in use is the evaluation for potential genotoxicity, as a screen for chemicals which might possess mutagenic or carcinogenic properties (Elliott, 1994a). This may be for selection of candidate inks or colorants, for hazard assessment for employee and customer safety, or for submission to regulatory bodies for notification.

The core principle underlying genotoxicity tests is whether the chemical interacts with and damages the DNA of a test organism. The current strategy accepted by genetic toxicologists is that an initial assessment is made using *in vitro* assays (the Ames test) which is designed to detect any intrinsic genotoxic activity. If clear evidence of genotoxicity is seen, then *in vivo* assays are carried out to see if this activity is expressed in the whole animal.

There have been a large number of false-positives found using the *in vitro* tests, i.e. chemicals which show genotoxicity *in vitro* but which are not genotoxic or carcinogenic *in vivo* (Elliot *et al.*, 1993). Thus, decisions on the toxicology of a compound cannot be based solely on the *in vitro* data alone. The *in vivo* assays now used are the mouse bone-marrow micronucleus assay and the rat-liver unscheduled DNA synthesis assay. These provide an effective screen for the detection of genotoxic carcinogens.

A number of aromatic amines show a strong positive response in the Ames test and 2-naphthylamine (**26**) has been shown to be a potent carcinogen. However, the introduction of water-solubilising sulphonic acid or carboxylic acid groups often renders these types of molecules inactive and 2-aminonaphthalene-1-sulphonic acid (**27**) is a perfectly safe chemical.

(26)   (27)

The dyes used in aqueous-based ink jet inks all contain an abundance of sulphonic acid and/or carboxylic acid substituents to give them their water solubility. As such, most of these dyes are safe and give negative responses in both the *in vitro* and *in vivo* tests.

Increases in the legislation covering the disposal of waste could have major implications in the ink jet printing industry. One aspect of this could be that disposable cartridges are returned to the manufacturer during delivery of replacements. Alternatively, printheads of the permanent type that can be easily refilled may be more prevalent in future printers.

The recycling of printed paper and board is now becoming widespread and problems have been encountered with 'de-inking' some of these newer technology inks. If ink jet printing significantly increases its market share, studies on 'de-inking' will become necessary. Novel colorants that are removed from paper by a simple process may emerge in the future.

5.12 Future of ink jet

During the last decade, non-impact printers have revolutionised the printing industry. Laser printers and ink jet printers have gained market share at the expense of impact printers such as dot-matrix. Having successfully displaced these impact printers, the next battle will be between the non-impact printers themselves. For monochrome (black) printing in the office, photocopiers and laser printers will probably still predominate. Their high print quality at reasonable cost will make them difficult to displace. However, with ink jet continually improving and near letter-quality text now achievable, it could capture a significant share of the lower end of this market. For full-colour imaging, ink jet should become the dominant technology. It is now capable of giving acceptable print quality at costs far lower than competing technologies. Improvements in speed and print resolution are likely to continue and new technologies such as the Microjet system could emerge. Page-wide arrays, probably of the piezo type, would give much faster speeds and would find use in the short-run printing area.

If progress continues at the present rate, the use of drop-on-demand ink jet printing will become widespread in the office environment, the major uses being in colour printers, fascimile printers and wide-format office plotters. The worldwide installed base of ink jet printers is expected to grow from 16 million units in 1994 to approximately 30 million units by 1998. If these trends are correct, the sales of ink jet printers relative to laser printers will increase from about 50% in 1994 to about 75% in 1998. The future of ink jet will undoubtedly be in economical full-colour printing with continually improved performance.

A recently announced technology, TonerJet, combines ink jet and laser-printer technology (Johnson and Larsson, 1993). It fires dry toner through a grid on to the substrate where it is fixed by heat fusion. Whether TonerJet becomes a major technology remains to be seen.

References

Elliott, B.M., Jackh, R. and Jung, R. (1993) Evaluation of the Genotoxicity of 4-Diethylamino-4'-nitroazobenzene and Seven Analogues. ICI Internal Report.

Elliot, B.M. (1994a) Genotoxicity Evaluation in Inks. *IS&T's Tenth International Congress on Advances in Non-Impact Printing Technologies*, New Orleans, pp. 443–445.

Elliott, P. (1994b) The Impact of Environmental Issues on the European Non-Impact Printing Industry. *IS&T's Tenth International Congress on Advances in Non-Impact Printing Technologies*, New Orleans, pp. 7–9.

Exxon (1984) European Patent 181 198A.

Gregory, P. (1991) Ink jet printing. In *High Technology Applications of Organic Colorants*, pp. 175–205. Plenum, New York.

Gregory, P. (1994) Choosing Colourants for Ink Jet Printing – Dyes vs. Pigments. The Truth. *IMI 3rd Annual Ink Jet Printing Workshop*, Cambridge, Massachussetts, 5–6th April.

Hammond, D.L. (1984) *Hewlett Packard Journal*, March, p. 44.

Hertz, G.H. and Heinzl, J. (1985) *Advances in Electronics and Electron Physics*, **65**, 91.

Howtek (1984) European Patent 18 7352A.

Johnson, J. and Larsson, O. (1993) Tonerjet. A Direct Printing Process. *IS&T's Ninth International Congress on Advances in Non-Impact Printing Technologies*, Yokohama, p. 509.

Kenyon, R.W. (1994) Dyes for Ink Jet Printing. *Innovations in Modern Colour Chemistry*, SCI, London, 27 April.

Lyne, M.B. (1986) *Journal of Imaging Technologies*, **12**(2), 80.

Provost, J. (1994) Ink jet printing on textiles. *Surface Coatings International*, **77**(1) 36–41.

Willis, M. (1991) *The Path to Page-Wide Ink Jet Arrays, BIS Ink Jet Conference*, Amsterdam, 21 March.

Work, R.A. (1994) Ink and Media Advances Needed for Improved Ink Jet Competitiveness. *IMI 3rd Annual Ink Jet Printing Workshop*, Cambridge, Massachusetts, 5–6th April.

6 Thermal printing

R. BRADBURY

6.1 Introduction

Thermal printing, also known as thermography, embraces the various reprographic technologies which utilise heat as the energy source. Most commonly, this heat is applied by either thermal heads or a heated stylus. However, other methods such as resistive ribbon technology may be utilised and there are increasing signs that infrared lasers are set to play a major role in future thermal printing technologies. There are two main areas of thermal printing: direct thermal printing and thermal transfer printing. Thermal transfer printing itself may be conveniently subdivided into dye-diffusion thermal transfer, D2T2, and thermal wax transfer (thermal melt transfer, thermoplastic transfer).

The evolution of thermal printing is depicted in Figure 6.1. Direct thermal printing is the most mature of the technologies, the origins of which can be traced back to the 1930s, although the form in which we recognise it today emerged in 1953. This remained the dominant thermal printing technology for over twenty years until the invention of thermal wax transfer during the mid-1970s. The development of this new technique offered several advantages over direct thermal printing: the need for specially coated papers was eliminated, archival storage was greatly improved and, perhaps most importantly, full-colour imaging became possible although direct thermal has since taken a further step forward with the development of 'Thermo-Autochrome' printing by Fuji Photo Film. Less than a decade later, the introduction of dye-diffusion thermal-transfer printing produced the next step change in thermal printing bringing the possibility of high quality photographic-like reproduction from electronically captured/generated images. This was quickly followed by the invention of electrothermal (resistive ribbon) technology by IBM which replaces the need for a heated stylus or thermal head by coating the material to be transferred on to a base film which is electrically resistive and can therefore be heated by applying electrical energy to the film.

As a general rule direct thermal printing represents the lower quality, lowest cost technology whilst D2T2 provides very high quality but at a higher price.

Although the initial concept of the Cycolor imaging process developed

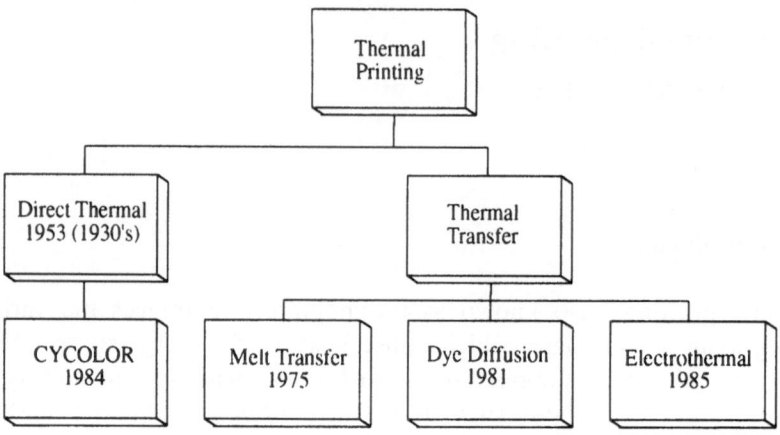

Figure 6.1 Evolution of thermal printing.

by the Mead Corporation was not as a thermal printing but as a pressure-sensitive system the technology is included in this overview since the chemistry utilised to generate the coloured image is very closely related to that of direct thermal printing. In addition, however, the process incorporates a photopolymerisation step in order to faciliate the production of full-colour images. This is in contrast to the usually monochromatic output from direct thermal printing.

6.2   Direct thermal printing

The first direct thermal-printing technique emerged in the 1930s and utilised a physical rather than a chemical change to produce the image. The method involved coating a wax on to either a black paper or black-coated base paper and then writing on this with a heated stylus. The heat melted the wax which flowed to the sides revealing the black base paper (Figure 6.2). Although not particularly sophisticated, such papers found use in

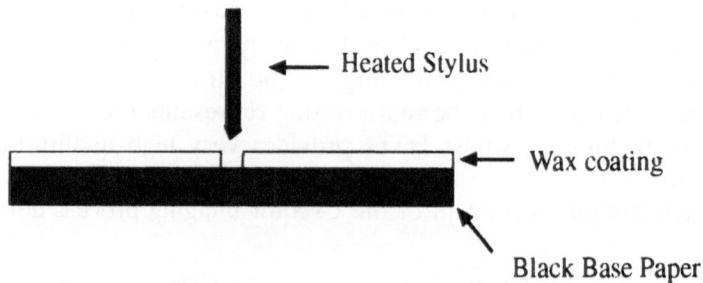

Figure 6.2 Early direct thermal printing.

medical and scientific applications, being suited, for example, to electro-cardiography where speed is far less important than simplicity.

Significant drawbacks with this method were the mechanical stability of the wax coating and also low thermal sensitivity. Additionally, the coated paper has a different feel from ordinary paper and the background is not completely white.

During the 1950s, Minnesota Mining and Manufacturing Company (3M) introduced Thermo-Fax paper which comprised a heat-sensitive coating applied to a base paper. This coating consisted of a metallic salt of an organic acid together with a reducing agent dispersed in a binder. The coated face of the paper was placed in contact with a document original and an infrared source used to apply heat through the document (Figure 6.3).

Duplication was effected by differential heating of the original causing a reduction of the metal salt in the heat-sensitive layer to leave a dark image. This technique gave a simple method for copy reproduction and rapidly captured a large share of the US market. Unfortunately for this particular technique, electrophotography (photocopying) (see Chapter 4) came to dominate the area of duplication and resulted in the demise of this system.

Further advances in thermal printing would have been greatly restricted without the ensuing developments in modern, solid-state, high-speed thermal printheads started during the 1960s. It was during this time that the NCR Corporation developed a low-cost thermal paper based on colour generation from the reaction of a colour former with a colourless acidic co-reactant. This paper was stable at temperatures up to 60°C and gave a blue image when subjected to temperatures in the 100–150°C range. This is the principle on which modern thermal papers are based. The construction of the thermal paper is broadly similar to that of the Thermo-Fax paper described above. The heat-sensitive layer, generally 5–10 μm in thickness, consists of fine dispersions of both the colour former and an acidic organic compound combined with, but held apart by, a binder (Figure 6.4). Typical binders are polyvinyl alcohol, carboxymethyl cellulose, hydroxyethyl

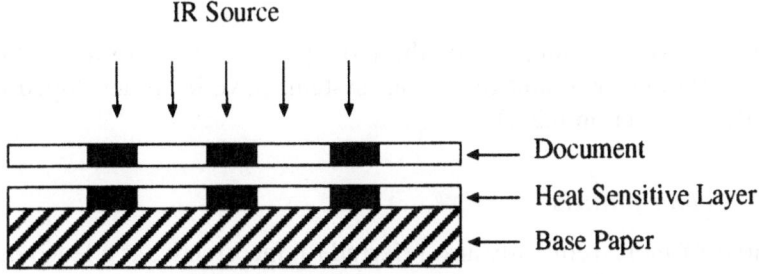

IR Source

Document

Heat Sensitive Layer

Base Paper

Figure 6.3 Thermo-Fax copying system.

← Heated Stylus

○ Colour former

△ Developer

Figure 6.4 Thermal paper construction.

cellulose, polyacrylamide and polyvinylbutyral used either singly or in combination.

Application of heat from a thermal head causes localised melting of the acidic component which brings it into contact with the colour former generating the colour. Crystal violet lactone (**1**) is typical of the colour formers and bisphenol A (**2**) of the acidic organic reagent. These two components can be used to illustrate the colour generation reaction (Equation 6.1).

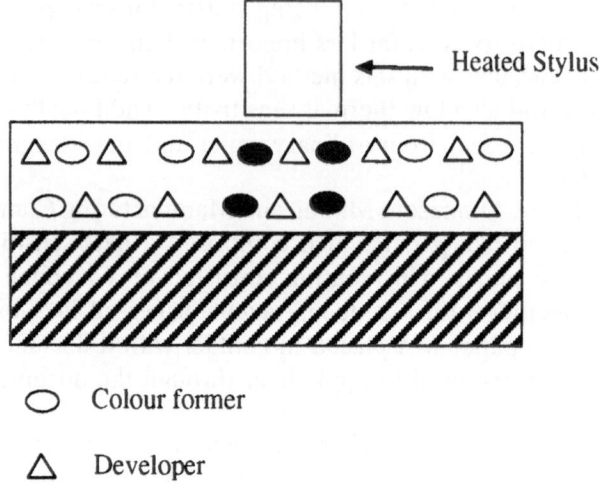

(**1**)　　　　　　　　　　　　　　(**2**)　　　　　　Violet　　　　(6.1)

Direct thermal printing is for the most part restricted to monochrome images although two- and full-colour systems have been developed more recently (see section 6.2.4).

6.2.1　Colour formers

A colour former, frequently also referred to, incorrectly, as a leuco dye, is an essentially colourless molecule which, on contact with a colour developer, generates an intense colour. For thermal printing the most important colour is black and this may be achieved either by using a mixture of colour formers such as an orange (e.g. **3**) and a blue, such as crystal violet lactone (**1**).

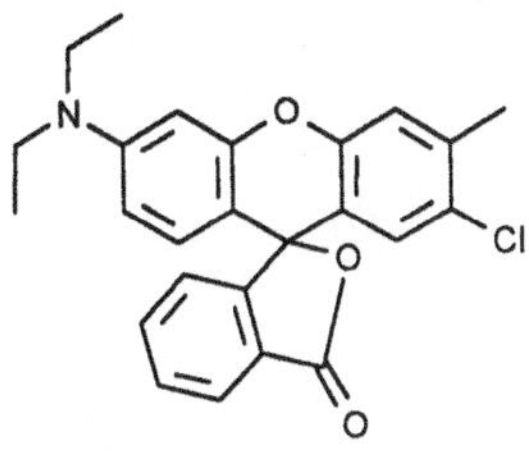

(3)

The most widely used black colour formers are, however, single molecules, referred to as 'one-dye blacks' developed during the 1970s by Yamamoto Chemical Corporation and based on fluoran structures. Reaction with the developer produces an intensely coloured cationic xanthene dye. The first one-dye black (**4** has been superseded by the related dyes (**5**) and (**6**) and despite intensive research over the intervening years these remain the most important thermal dyes in Japan.

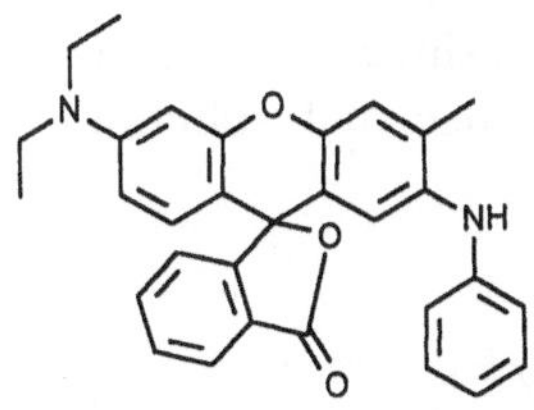

(4)

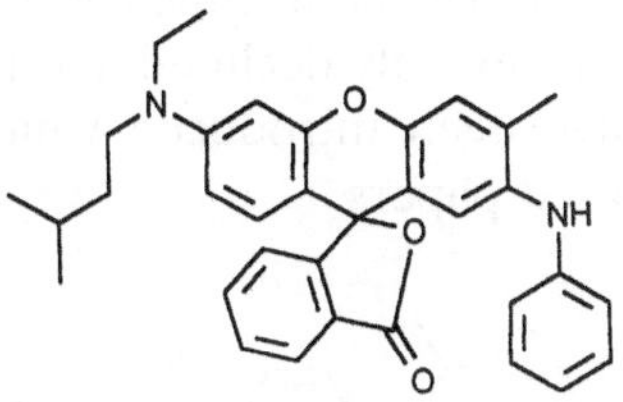

(5)

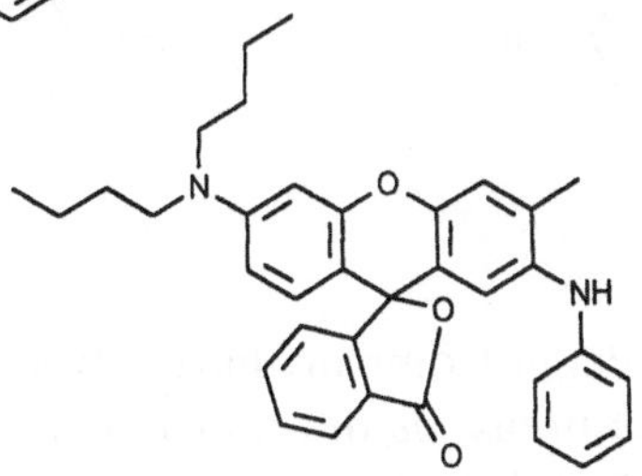

(6)

A range of colour formers are available which may be used either singly or in combination to generate images of a number of different colours. For instance, in addition to the blue and orange colour formers already mentioned, red colour formers are known (e.g. **7**).

(7)

6.2.2 Developers

The developers used are usually acidic solids, particularly phenols. The most common amongst these is bisphenol A (2) which accounts for over 70% of the Japanese market. One of the major attractions of bisphenol A is its relatively low cost. It is also readily available, being synthesised on a large scale for use in the manufacture of polymers. However, the colouring sensitivity is somewhat low which means that a relatively large energy input is required to achieve high-image densities. 4-Hydroxy-4'-iso-propoxy-diphenylsulphone (8) also finds significant usage in the production of high-grade thermal papers where image stability is required. Use of (8) imparts better moisture resistance and longer image lifetimes but at the penalty of higher cost. Benzyl 4-hydroxybenzoate (9) was once an important developer, being much more sensitive than bisphenol A although usage has now severely declined due to the higher cost and also since new sensitisers have been introduced which enhance the performance of bisphenol A-based papers.

(8)   (9)

Recently, the New Oji Paper Company have introduced a new group of compounds, the sulphonyl ureas, as developers (Takahashi *et al.*, 1994). Both mono- and bis-sulphonyl ureas are found to be effective with the monosulphonyl ureas, e.g. N-*p*-toluenesulphonyl-N'-phenylurea (10) claimed to give a more stable image than that formed using bisphenol A. By appropriate selection of sensitiser, the sensitivity of the paper can be adjusted to almost the same level as high-sensitivity papers based on bisphenol A. However, the problem of poor image stability in the presence of plasticisers remains. This is overcome by the use of bis-sulphonylureas (e.g. **11**) which exhibit excellent properties as developers and also produce very stable images.

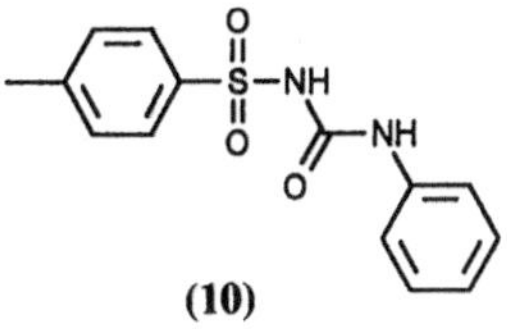

(10)

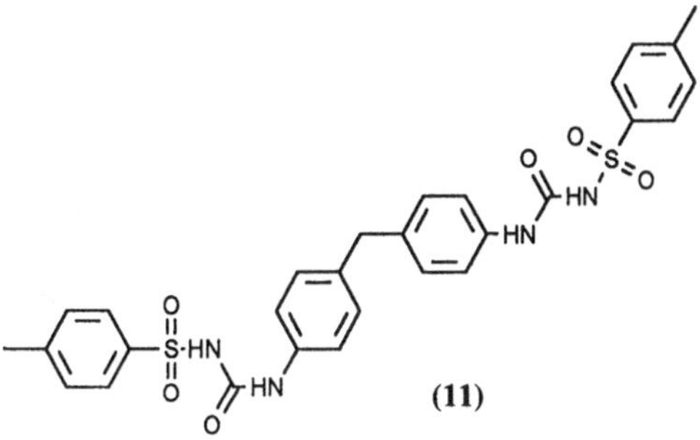

(11)

6.2.3 Sensitisers

An alternative means of reducing the energy input required for the generation of high optical density is to incorporate a third component, a sensitiser, into the thermal paper. The function of a sensitiser is to melt during printing and in doing so act as a solvent thereby facilitating contact between the colour former and the developer. In this way, the performance of bisphenol A-based systems can be improved avoiding the need for much more expensive developers. Sensitisers account for the largest production volume of thermal imaging chemicals with benzyl 2-naphthyl ether (**12**) the most widely used of these. Others include dimethyl terephthalate (**13**), 2-chloropropionanilide (**14**), 4-benzyldiphenyl (**15**), 1,2-di-(3-methylphenoxy)ethane (**16**), dibenzyl oxalate (**17**) and m-terphenyl (**18**).

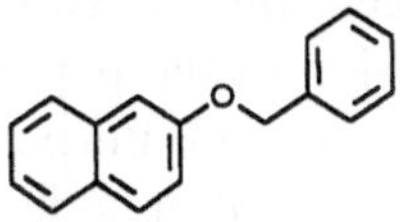

(12)

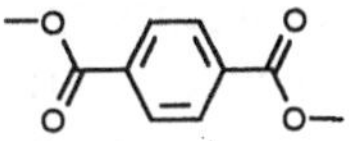

(13)

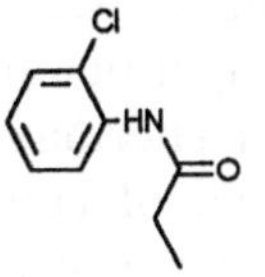

(14)

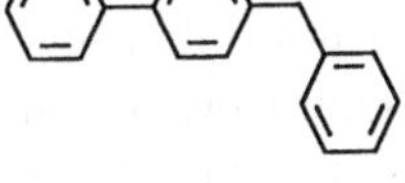

(15)

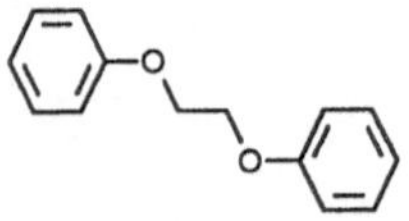

(16)

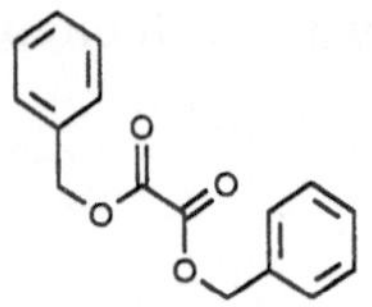

(17)

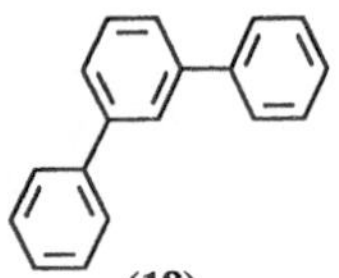

(18)

By reducing the energy input requirements, sensitisers contribute to thermal head life and have enabled portable, battery-powered printers to be developed.

Images produced using thermal papers tend to display poor lighfastness. To minimise this problem, ultraviolet absorbers such as (19) and (20) are added to the heat-sensitive layer. Alternatively, polymer overcoats may be applied to the paper to protect the image.

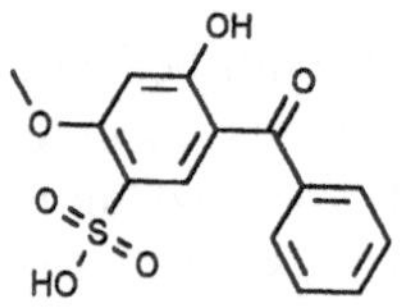

(19)

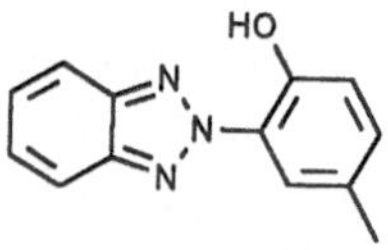

(20)

Because the thermal head comes into direct contact with the paper, antistick agents are necessary to prevent adhesion of the paper to the hot thermal heads. A range of materials are available for this purpose, e.g. zinc stearate, various fluoropolymers or fine-particle size inorganic compounds such as calcium carbonate or talc.

A number of disadvantages of thermal paper and the steps which may be taken to overcome them have already been mentioned. In addition, the paper has a high surface gloss which makes reading difficult and the smooth surface not only imparts a strange feel to the paper but also makes the paper difficult to write on. In response to this Plain Thermax has been developed (Watanabe, 1993). This paper has the feel of conventional paper and has a matt surface provided by an overcoat layer which can be written on easily. This overcoat also provides abrasion resistance and protects the image from contamination. The binder used improves water resistance and prevents the sticky feel when touched by a damp hand. High sensitivity is achieved by including an oil-absorbing pigment in the base coat which has an insulating effect reducing the energy loss from the thermal head. Plain paper texture is achieved by using a stiff base paper of 70–90 μm thickness.

6.2.4 Multicolour thermal paper

Although monochrome imaging dominates direct thermal printing, two systems for producing images in two colours have been proposed. One of these consists of a paper having two separate thermal layers each of which develops a different colour. The upper layer melts at a lower temperature providing the first colour at low-energy inputs. The second colour is produced at higher energy inputs when both layers melt together, thereby mixing the two dyes. Results were not entirely satisfactory and so the second method was developed by Kanzaki Paper. This system employs an erasing layer between the two colour layers. The upper layer develops colour at low-energy inputs but when higher energy is applied in order to generate the colour of the lower layer, the intermediate erasing layer is brought into contact with the first colour and destroys it. In this way, two distinct colours can be printed (Figure 6.5).

Recently a full-colour direct thermal printing method known as 'Thermo-Autochrome' has been developed by Fuji Photo Film (Igarashi, 1994). The method uses three separate colour layers – yellow, magenta and cyan – protected by a fourth, heat-resistant layer. However, only the cyan layer utilises the colour former/phenolic developer reaction which has been discussed in some detail above. The yellow and magenta colour-generating layers both utilise azo dye coupling reactions followed by photolytic decomposition of residual diazonium salt by ultraviolet light of specific wavelengths. Thus, 'Thermo-Autochrome' is a combination of the direct thermal and diazo-type technologies. In order to be able to use reactive

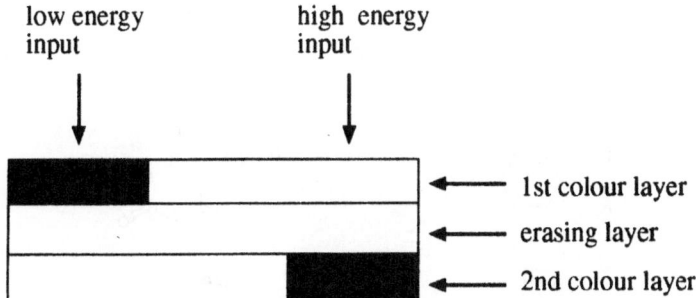

Figure 6.5 Multicolour thermal paper with erasing layer.

diazonium salts in a practical system it was necessary to find a means of preventing contact between the diazonium salt and the coupler until thermal imaging is initiated. Fuji Photo Film therefore also developed 'Micro Isolation' technology in which an oil-based solution of the required diazonium salt is microencapsulated. The walls of these microcapsules are made from poly(urea/urethane) which, when heated above its glass-transition temperature, becomes permeable allowing the coupler together with an organic base to enter the capsules and react with the diazonium salt thus generating the colour. In addition to giving control over the colour generation, the encapsulation process confers good shelf-life such that after two years storage at room temperature >90% of the diazonium salt remained intact. The full-colour imaging process occurs in five steps. The yellow colour is generated first by low thermal energy input from a thermal head. The entire print is then irradicated with light at 420 nm which selectively decomposes the residual yellow colour forming diazonium salt. The magenta colour is then generated following a higher energy input and the residual magenta colour forming diazonium salt decomposed by ultraviolet light at 365 nm. The cyan colour is then developed by higher energy input utilising 'conventional' colour former/developer chemistry. The result is a full-colour image.

The water solubility of the diazonium salts and the cyan colour former used are important factors in determining the stability of the thermal paper. If this is too high, leakage of the colour precursors through the capsule walls will take place and premature colour generation will occur. The diazonium salts are therefore derived from hydrophobic counter ions such as PF_6^-, BF_4^-, in addition to containing hydrophobic substituents. Typical structures are (21–24).

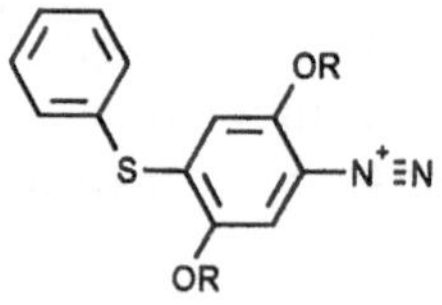

(21) 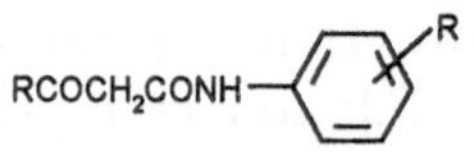   (22)

Yellow

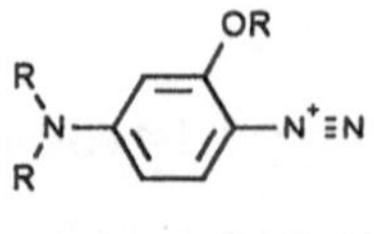

(23) 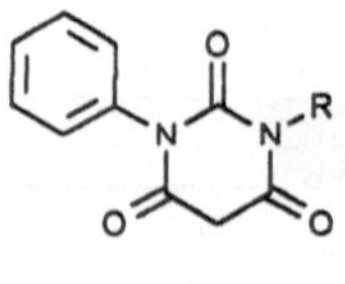   (24)

Magenta

It is claimed that the 'Thermo-Autochrome' system produces high-resolution images which are sufficiently stable for most applications although it is noted that over time there will be a slight increase in background density due to the degradation of the colour precursors used. Quality is said to be comparable to D2T2 but energy consumption is higher and the lightfastness properties are inferior. However, 'Thermo-Autochrome' is a very recent development and efforts are being made to overcome the deficiencies.

6.2.5 Direct thermal markets

The thermal paper industry is highly competitive and has been led for the most part by Japanese technology. Although a few European companies are active in the area, US subsidiaries and Japanese licencees are more common. Much of the growth of thermal printing over the last twenty years has been driven by the proliferation of fax machines. The current trend towards plain-paper fax machines will no doubt begin to dictate progress in the future. Prior to 1989, European demand for thermal paper far exceeded supply, the balance being met by imports from Japan. However, in the period 1991–92, European capacity tripled as companies such as Ricoh and Kanzan opened up manufacturing facilities and by mid-1992, supply was almost twice the demand. Together with Arjo-Wiggins, these two companies account for over half of the European production.

6.3 Mead Cycolor process

Although not a direct thermal printing process, the Mead Cycolor process is included here since the colour-generation step involves the same colour former and developer technology as direct thermal printing, the two being brought into contact by the application of pressure rather than heat. More importantly, however, the Cycolor technology was designed to allow full-colour, continuous-tone, imaging as opposed to the essentially monochrome imaging available at the time from direct thermal printing. This was achieved by microencapsulating the colourless dye precursor (colour former) and a photoinitiator together as a solution in a liquid monomer. The Mead Corporation termed these microcapsules 'Cyliths' and the image reproduction process 'Cylithography'. Three photoinitiators are required and selected such that they are sensitive to blue (380–480 nm), green and red light. Indeed, the discovery of efficient, visible light-sensitive photo-initiators was a key step in the development of a commercial imaging system as earlier systems had been restricted to exposures between 350 and 490 nm requiring artificial colour separation. By incorporating the photo-initiators separately into the monomer solutions of the yellow, magenta and cyan colour formers, respectively, it is possible to selectively photopolymerise the liquid monomer resulting in solidification of the encapsulated solution which can no longer rupture when pressure is applied during transfer to the receiver sheet. In contrast, the capsules which have not undergone the photopolymerisation step do rupture allowing the colour former to come into contact with the developer and generate the colour. Thus, exposure of the pressure-sensitive layer to blue light prevents development of yellow by polymerising the yellow colour-former capsules. Similarly, exposure to green light polymerises the magenta colour-former capsules and red light the cyan colour-former capsules. In the absence of any photoexposure, all capsules are ruptured giving a black image and if the sheet is exposed to white light then all capsules are polymerised and white results. The foregoing description deals with the extreme situation where all the monomeric solution is polymerised. However, by controlling the polymerisation step partial polymerisation can be effected leaving part of the colour-former solution as a liquid which is then able to come into contact with the developer on rupturing under pressure, thus providing continuous tone capabilities. The process has been described in more detail by Rastogi and Wright (1989) of Mead Imaging. In order to overcome oxygen inhibition of the photopolymerisation step, additives which can consume oxygen in free-radical chain processes are included. Such compounds include 4-acetyl-*N*,*N*-dimethylaniline, 2,6-diisopropylaniline and 4-*t*-butylaniline (European Patent 223 587 Mead Corporation). Suitable monomers include trimethylolpropane triacrylate (TMPTA) and tetraethylene glycol di-acrylate (TEGDA).

6.3.1 Photoinitiators

Two classes of photoinitiators have been used to realise the goal of visible light sensitivity. The first was a conventional ketone/amine initiator system whilst the second was based on cationic cyanine dyes which were capable of generating free radicals on photolysis. Mead developed cyanine borate salts which are efficient initiators for the polymerisation of acrylates and are visible light sensitive. The mechanism by which these cyanine borates generate free radicals is not fully understood.

The microcapsules are made by dispersing a solution comprising the appropriate colour former and photoinitiator in a monomer, in water containing surfactants and appropriate viscosity controlling agents whose function is to control the size of the microcapsules. The capsule walls are former from water-soluble, thermally polymerisable monomers. On addition of a catalyst, the water-soluble monomer slowly polymerises and begins to separate from the solution encapsulating the hydrophobic monomer solution in the process. Further polymerisation and cross-linking of the hydrophilic monomer gives a discrete capsule wall. The process is repeated for each of the colour former/photoinitiator combinations. It is important that the capsule wall-forming material is transparent to the radiation being used and a preferred material is a urea–resorcinol–formaldehyde resin in which the resorcinol component enhances the oleophilicity of the polymer.

Several applications of the technology have been developed. One of the simplest is that of a 35 mm slide printer. In this application the 35 mm original is projected on to the photosensitive sheet which is then passed between pressure rollers rupturing the non- or partially polymerised microcapsules giving a full-colour print of the original. Alternatively, a copy system may be constructed by illuminating an original image with a tungsten halogen light source and projecting the image on to the paper. The image is again developed by application of pressure.

Despite offering several advantages, such as high-speed copies of high resolution at relatively low cost, the Mead process still has limitations not least of which are the archival storage properties. The colour formers and developers used are the same or similar to those employed in other thermal and pressure-sensitive printing systems, all of which have limited light-fastness. Unfortunately for Mead who invested heavily in this technology forming a new company, Mead Imaging, early optimism surrounding potential applications has proved unfounded and the system has been unable to compete with alternative colour-imaging technologies such as electrophotography (see Chapter 4) and ink jet (see Chapter 5).

6.4 Dye-diffusion thermal transfer (D2T2)

The invention of the Mavica still video camera and associated Mavigraph printer by Sony in 1981 heralded the birth of a new technology for the capture and reproduction of colour images. Despite considerable research, an overview published a few years later (Fox, 1988) of the subsequent development of electronic cameras indicates that the attainment of photographic quality images using still video cameras is still some way off. However, the inevitable desire for hard copy of electronically captured images from a variety of other sources has provided a driving force for the development of a suitable printing technology. Because the amount of dye transferred is in proportion to the heat applied dye-diffusion thermal transfer (D2T2) printing has the advantage of continuous tone reproduction and in recent years has become recognised as the non-impact printing (NIP) process capable of producing images of near photographic quality and, although the advent of electronic photography has been delayed for reasons referred to above, D2T2 has found major uses in security, medical, scientific and novelty markets. Several of the large US (e.g. Eastman

Kodak) and Japanese (Hitachi, Mitsubishi, etc.) companies currently offer printers and associated consumables, dyesheet cassette and receiver sheets, to the marketplace. Others (e.g. ICI) offer consumables for a variety of printers. More recently, Kodak has introduced the ColorEase PS colour printer designed to produce colour transparencies and prints of near photographic quality for business communications.

D2T2 printing bears some similarity to melt-transfer printing in that both systems utilise a colour donor ribbon. However, there are fundamental differences not least of which is that, in contrast to wax transfer, the colorants used in D2T2 are required to penetrate into the receiver sheet leaving the binder behind on the dyesheet. This transfer of the dye from the dyesheet and into the receiver sheet is now believed to proceed via a diffusion process and hence precludes the use of pigments such as those used in melt-transfer printing. Accordingly, the dyes should be thermally stable and diffuse readily in proportion to the heat applied. Although various means could potentially be used to supply the necessary heat energy, current commercially available printers employ thermal head technology. More recently, strong indications are emerging in the patent literature that lasers will play an important role in future generations of printers.

6.4.1   Dyesheet (dye donor sheet)

The fundamental purpose of the dyesheet is to retain the dye as a stable, homogeneous layer until transfer is effected by the thermal head. Meeting this apparently simple demand necessitates the composition of the dyesheet to be somewhat more complex than may appear on initial inspection. The construction of a typical dyesheet is shown in Figure 6.6.

The most commonly used support film is 4.5–6 μm thick polyester chosen because of its thermal conductivity, tensile strength, smoothness and relative cheapness. One major problem to be overcome is that the

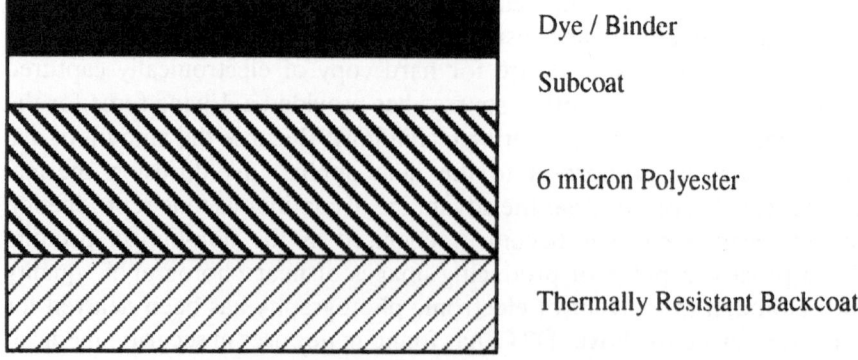

Figure 6.6 Typical dyesheet construction.

thermal heads can operate at temperatures up to 400°C which are well above the softening temperature of polyester. In order to alleviate this problem, the side of the polyester from which the heat is applied is coated with a thin, thermally stable backcoat. A considerable amount of research has been undertaken in order to arrive at suitable compositions for the backcoat since, in addition to demonstrating the necessary thermal stability, the coating has to have appropriate frictional properties to allow smooth passage of the dyesheet over the thermal head and must not cause any abrasive damage during use. As will be appreciated from Figure 6.6 when the complete dyesheet is spooled the backcoat comes into contact with the dyecoat and it is therefore important that none of the materials used to formulate the backcoat results in any deterioration of the dyecoat.

Below the dyecoat and on the opposite face of the polyester to the backcoat, a further thin coating, the subcoat, is applied. The primary purpose of this layer is to promote adhesion of the dyecoat although it may also prevent unwanted back diffusion of D2T2 dyes into the base film. Why this should be necessary becomes clearer on consideration of both the dyes themselves and also the usual receiver sheet construction (see later).

The dyecoat is generally prepared by coating a solution of dye and a binder, possibly with other additives, on top of the subcoat and is frequently accomplished using gravure coating technology in order to give the high-quality coating which is necessary to produce prints of photographic-like quality. The resultant dyesheet normally comprises a trichromat of yellow, magenta and cyan dyes coated as sequential panels of at least the same area as the print size. Depending on the printing application, a black dye panel may be incorporated either as a fourth panel or in certain instances as the only colour. Each trichromat repeat unit is interrupted by some form of registration mark which enables the printer electronics to determine the position of the dyesheet in the printer (Figure 6.7). Factors governing the choice of dyes are described in section 6.4.3.

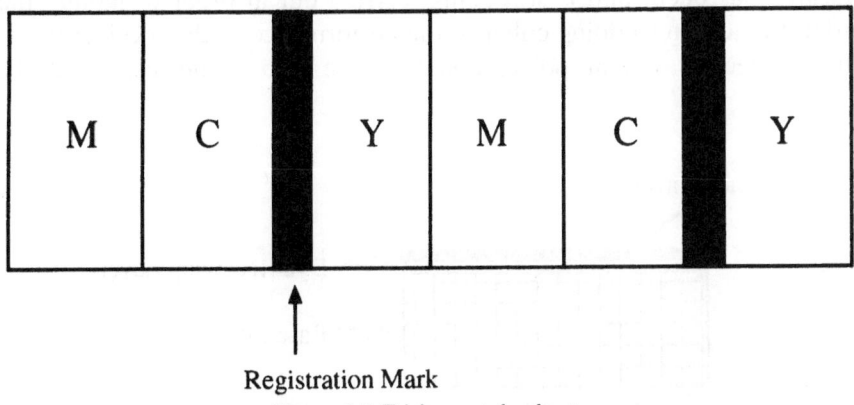

Registration Mark

Figure 6.7 Trichromat dyesheet.

The binder may be selected from a variety of materials, examples of which include cellulose derivatives such as ethylhydroxyethylcellulose (EHEC) and vinyl derivatives such as polyvinyl butyral (PVB). Mixtures of binder polymers and co-polymers can also be used. Once formed, the dyecoat must remain as an amorphous layer for a prolonged period to provide a dyesheet of adequate shelf-life. Crystallisation of the dye or other heterogeneity such as dust particles present in the dyesheet leads to a deterioration in print quality.

When all the above coatings have been applied to the polyester base film the dyesheet is spooled and supplied as a cassette, usually containing sufficient repeat units for fifty or a hundred prints.

6.4.2   Receiver sheet

The receiver sheet consists of three major components (Figure 6.8) – the base film, dye receiving layer and a release layer – and may be constructed by a number of techniques. Solvent coating is generally preferred as it permits a high degree of flexibility both in the construction of the receiver sheet and in the choice of additives which may be included, particularly in the dye receptive layer.

The dominant role of the base film is to provide dimensional stability and acceptable 'feel' to the receiver sheet. Both paper and white polyester are commonly used for this purpose and this makes up the bulk of the total receiver sheet thickness which is typically 150 µm. An alternative receiver sheet construction utilises a laminate structure wherein a paper base is sandwiched between two layers of white pigmented polypropylene. In the case of prints generated for use as overhead transparencies, then the white base film is replaced with clear polyester.

The dye-receiving layer has to be a clear polymer receptive to the dyes used in D2T2. As the dyes are mostly of the solvent or disperse types, then polyester receiver layers are frequently encountered and extensive research has been, and is still being, carried out in order to modify the properties and in so doing enhance the performance of the receiver layer. High molecular weight polycarbonates have also found use in D2T2

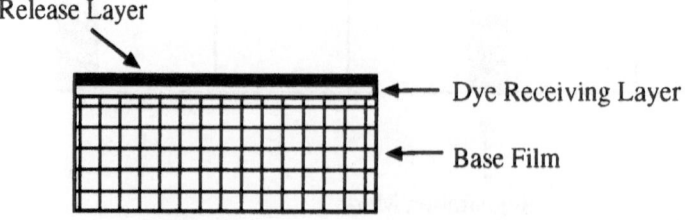

Figure 6.8 Receiver sheet construction.

(Eastman Kodak European Patent 227 094). Inclusion of a linear polyester or polycaprolactone to such a receiver layer has been claimed to improve lightfastness (Eastman Kodak European Patent 228 066). The thickness of the dye-receiving layer is typically <10 μm and in order to improve the lightfastness of the prints the incorporation of additives such as ultraviolet absorbers has been proposed.

In order to produce a D2T2 print, the dyesheet and the receiver sheet are held together against the thermal head under pressure from a platen roller. Application of heat from the thermal head causes diffusion of the dye from the dyesheet and into the receiver layer to form the image. This process can lead to fusion between the dyesheet and receiver sheet and to prevent this, a release layer is required. The release layer can be incorporated into the dyesheet (Eastman Kodak European Patent 227 092) but it is most commonly encountered as a component of the receiver sheet. This layer is introduced either by a separate coating step directly on to the surface of the receiver or by including the release agent in the receptive layer formulation. A range of release agents have been proposed for this purpose and include fluorinated surfactants, silicones and waxes.

6.4.3 Dye-diffusion thermal-transfer dyes

Of all the contributing components to D2T2 technology, the dyes are arguably one of the most crucial, particularly as it is the colour in the final image which is the feature immediately most apparent to the end user. With the continuing development of this technology there is a need to fulfil ever more stringent demands. Meeting these demands provides a challenge to the dye chemist and has prompted a high level of activity in this area. Not surprisingly, for the first few years following Sony's inventions, much of this was provided by Japanese companies. From about the mid-1980s, this has been supplemented by several European and US companies. Virtually all the published information regarding D2T2 dyes currently resides in the patent literature although an overview of D2T2 dyes has recently been published (Bradbury, 1995) which supplements two earlier Japanese reviews (Hashimoto, 1988; Murata et al., 1989). This is not particularly surprising since D2T2 is still far from being a mature technology and intellectual property is a priority for the companies working in the area. Some idea of the level of interest in this field can be gained from the fact that in the period from 1985 to 1993 several hundred patent applications have been published regarding dyes for D2T2.

In order to be suitable for use in D2T2 printing, dyes should meet many requirements:

- They should be of appropriate hue. For photographic-like applications this means that the three substractive primaries yellow (Y), magenta (M) and cyan (C) are required although the precise hue of any one of these dyes is to some extent dependent on the hue of the others. In different applications, dyes of other hues may be appropriate. Black is normally achieved by overprinting each of the primaries; however, a fourth panel on the dye ribbon comprised of a mixture of dyes which when printed gives a black image has been used.

- The dyes need to be bright having minimal unwanted spectral absorptions. It is desirable that a magenta dye, for instance, only absorbs green and not red or blue light.
- Tinctorial strength should be high to facilitate the attainment of the high optical densities required for high-quality prints.
- Once transferred, the dye must exhibit good print stability and should show no tendency to migrate out of, or within, the receiver sheet, both of which would lead to a deterioration in image quality. The prints must also be stable to handling and to the effects of light. Indeed, high lightfastness is probably the single most difficult aspect of print stability to achieve. The reasons for this become apparent on considering the environment in which a printed D2T2 dye is expected to exist. Although the total thickness of the receiver sheet is usually several tens of microns the actual dye receiving layer thickness is in single-figure microns and the dye itself resides in the top few microns of this. It is therefore readily exposed to the effects of light and the atmosphere which in combination can lead to photodegradation of the dye. This is in sharp contrast to the situation in textile applications where the dye is present homogeneously throughout a fibre, a hundred or more microns thick. Fading of the first few micons in the fibre would have virtually no discernible effect whereas in a D2T2 print, the effect is catastrophic.
- Last, but not least, toxicology is an important issue and dyes should be chosen with a view to safety both in manufacture and in use.

The foregoing is by no means an exhaustive list of the required properties but it does contain many of the more important features applicable to D2T2 dyes. The exact specification of, and relative emphasis placed upon, any of the above (or additional) criteria will depend on the end use envisaged for the final D2T2 image. The difficulties in maximising these properties in any single dye are readily apparent and some degree of compromise is obviously unavoidable.

With the exception of a few structures disclosed for the purpose of providing comparative data in patent applications, the identities of the dyes contained in commercial D2T2 consumables have not been openly disclosed. These patent applications, many of which are not yet granted, contain an immense number of possible dye structures of which only a very small proportion will so far have found commercial utility. In fact, the vast majority probably never will. The dyes contained in the remainder of this section are chosen to be illustrative of the area rather than a definitive list of those which have found use in existing products.

6.4.3.1 Yellow dyes. Although yellow anthraquinone dyes demonstrating high lightfastness have been proposed, e.g. (25) (3M, US Patent 5 061 678) the major interest appears to have been in dyes of the azo and methine classes, both of which are inherently tinctorially stronger and synthetically more versatile. While dyes of these types generally provide prints of high optical density, the attainment of high lightfastness has proved more difficult, thus providing a focal point for research. Although the pyridone-derived dyes have been shown to exist predominantly in the hydrazone tautomer, they are most commonly referred to as azopyridones and are synthesised by conventional azo dye chemistry; accordingly, they are considered here under this classification. One dye of this class (26) has been disclosed (Agfa-Gevaert European Patent 432 314) as having been used commercially.

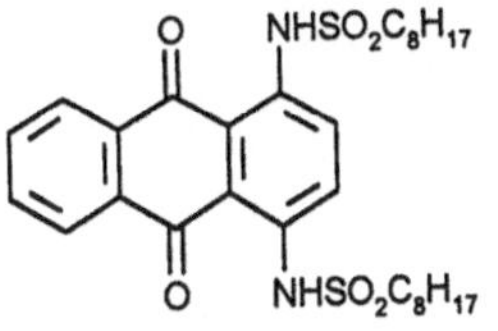

(25) 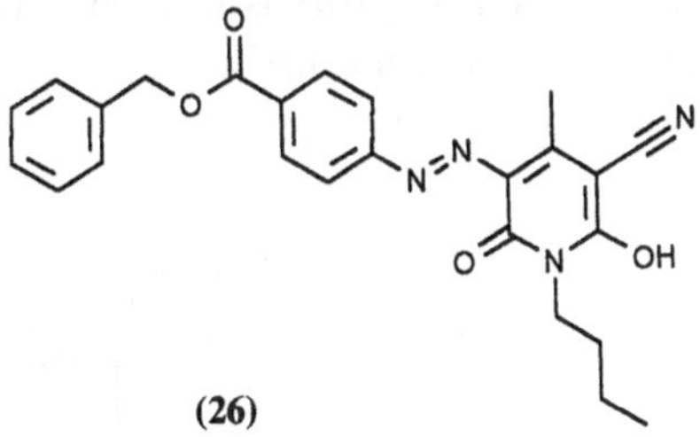   (26)

A great deal of research has been carried out on this class of dyes and a number of structural features have been found to influence the lightfastness properties. For example, it has been claimed that the use of 3-fluoroaniline as diazo component e.g. (27) confers increased lightfastness (Mitsubishi Kasei European Patent 442 466).

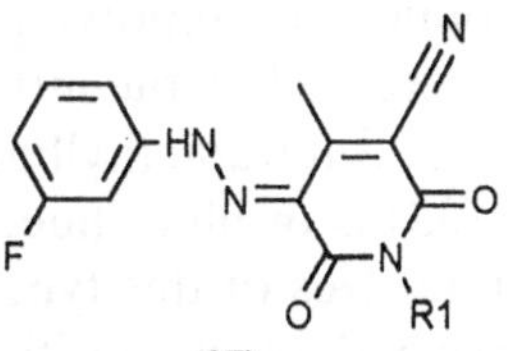

(27)

Many methine dyes have been proposed as yellows for D2T2. The dye (28) obtained by condensation of an aldehyde with malononitrile, is apparently used in a commercial product.

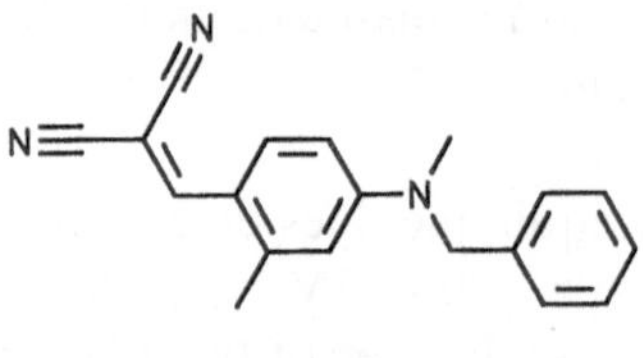

(28)

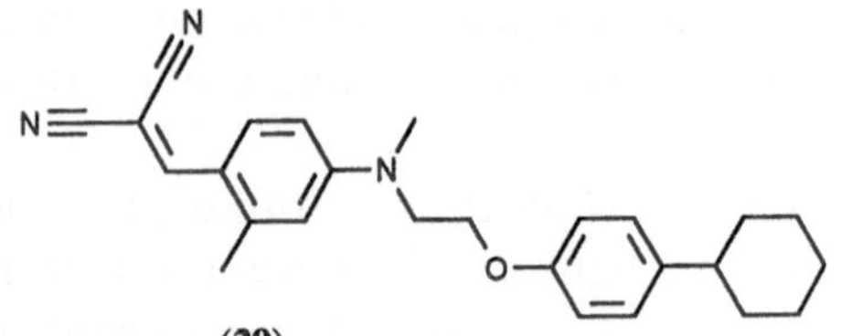

(29)

With dyes of this type, the print storage stability may be improved by the inclusion of *para*-substituted phenoxy groups and in this context the cyclohexyl group appears to have been used to good effect, e.g. (29) (Sumitomo Chemical Corporation US Patent 4 833 123). Improvements in lightfastness have been obtained with thiazole-derived methines in which the electron-accepting group is a 3-dialkylaminopyrazolin-5-one, e.g. (30). It appears that use of this type of electron-accepting group also minimises the problem of catalytic fading encountered when some methine yellows are utilised in conjunction with certain cyan dyes (Eastman Kodak European Patent 332 924). A further attraction in employing such heterocyclic acceptors is that they provide an alternative to the use of the highly toxic malononitrile.

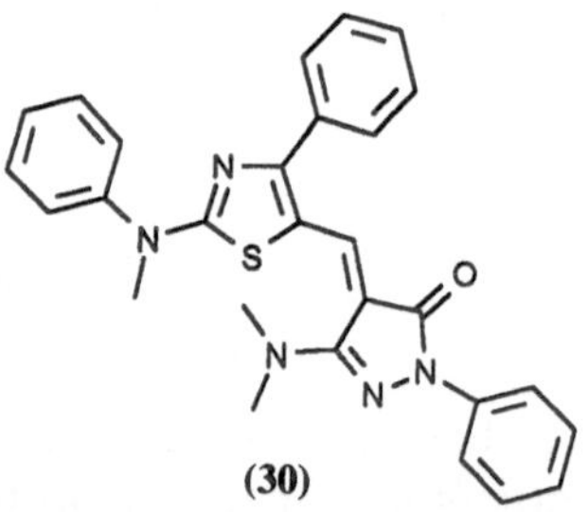

(30)

6.4.3.2 *Magenta dyes.*   In the magenta hue area, anthraquinone dyes, e.g. (31), generally exhibit good lightfastness properties and are readily synthesised from available intermediates. Despite this, the overall D2T2 performance of these dyes has been less than optimal and additional functionality has been introduced to improve properties such as solubility and hue. Solubility can be increased by the introduction of sterically bulky alkyl substituents such as 1,1,3,3-tetramethylbutyl into the phenoxy ring of (31) to give (32), whereas (33) is of bluer hue.

It has also been found that dyes of this type containing groups such as alkoxy, halogen and hydroxy in the *meta*-position of the phenoxy ring have higher solubility than their *ortho*- or *para*-isomers. The use of mixtures of dyes such as (31) and (33) allows the attainment of bluer hues with the added benefit of a synergystic improvement in solubility. The synthetic versatility of azo dyes has been exploited to good effect with dyes such as (34) being proposed for use as magentas. Although these dyes have good

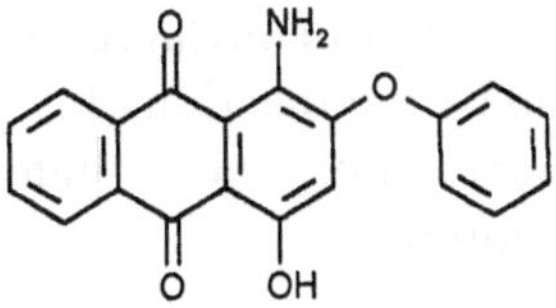

(31)

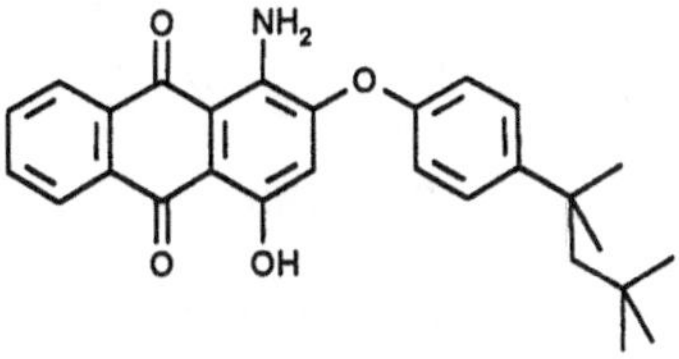

(32)

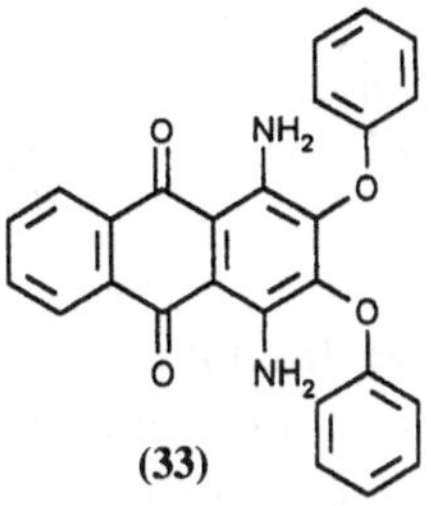

(33)

lightfastness properties, solubility in ketonic solvents, e.g. 2-butanone, can be a problem. This has been overome by replacing the alkylcarbonylamino group in (34) with an arylcarbonylamino group e.g. (35). Lightfastness properties of these dyes may be enhanced by the introduction of ester substituents into the N-alkyl groups of the coupling component.

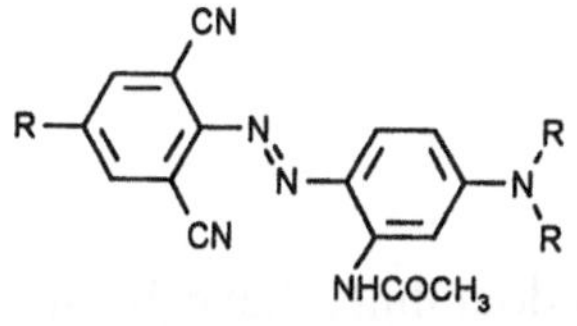

(34)

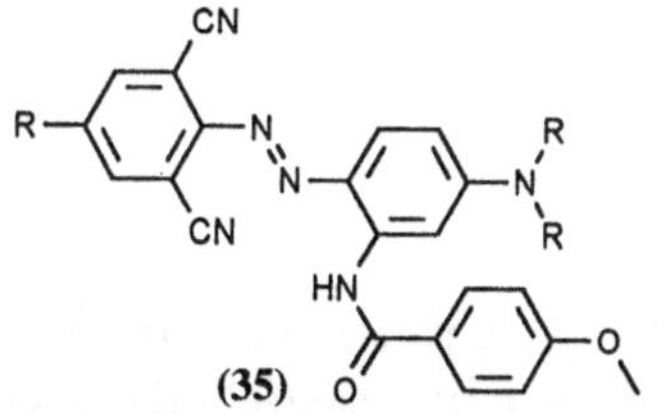

(35)

Azo dyes derived from heterocyclic diazo components have been a fruitful area of research for magenta dyes. In particular, 5-amino-4-cyano-3-methylisothiazole gives dyes of good magenta hue and D2T2 performance (ICI European Patent 216 413 and Eastman Kodak US Patent 4 698 561), e.g. (36) and (37). It has been found that the lightfastness of a number of azo dyes may be improved by branching of the N-alkyl substituents in the coupler (ICI WO94/08797).

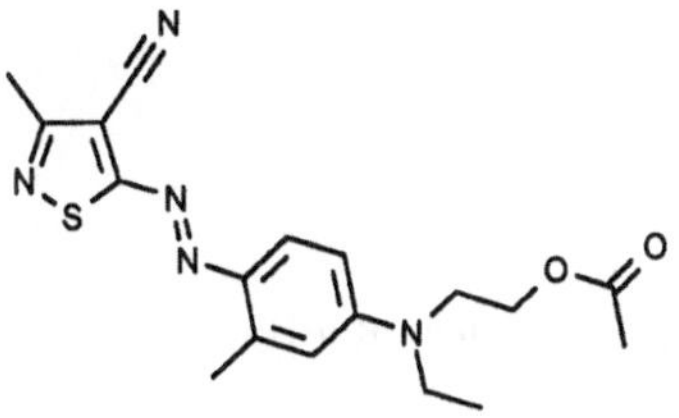

(36)

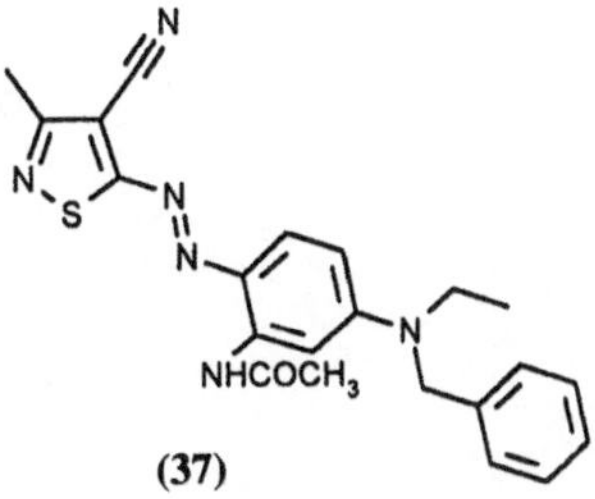

(37)

A range of very bright magenta dyes which give high print optical densities is provided by the tricyanovinylarylamines (**38**). These dyes may be synthesised in good yield by the reaction of N,N-dialkylanilines with the highly reactive, but unfortunately also highly toxic and expensive, tetracyanoethylene (Equation 6.2).

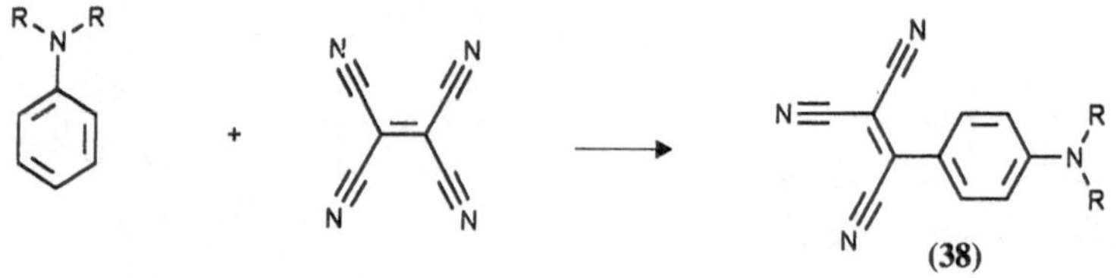

(6.2)

(**38**)

In use, the major deficiency of these dyes is limited lightfastness and N-alkyl chain branching has also been found to be effective in increasing the lightfastness of these dyes (Agfa Gevaert European Patent 593 817). An alternative approach to a higher lightfastness methine magenta is illustrated by (**39**).

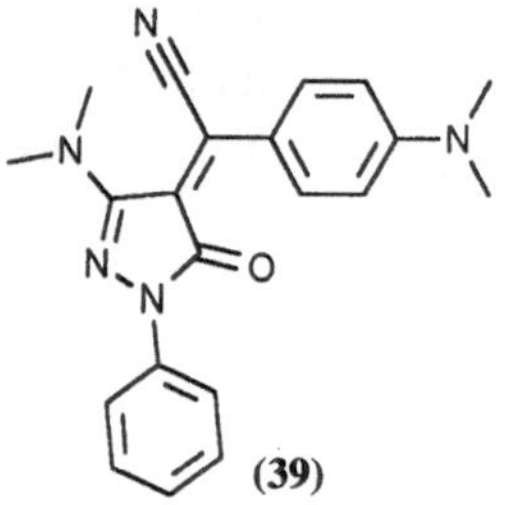

(**39**)

6.4.3.3 Cyan dyes. In the cyan area, the indoanilines have been the most heavily researched class of dyes. This level of interest is not surprising as they are the classic photographic cyan dyes. As with most D2T2 dyes, the major problems which have had to be solved have been in the area of print stability and particularly lightfastness. Some of the earliest examples gave only moderate lightfastness and displayed a marked dependence on the nature of the receiver polymer. Mitsubishi found that (**40**) gave much poorer light fastness in a PVC than in a polyester based receiver whereas (**41**) had much better light fastness and behaved similarly in both polymers.

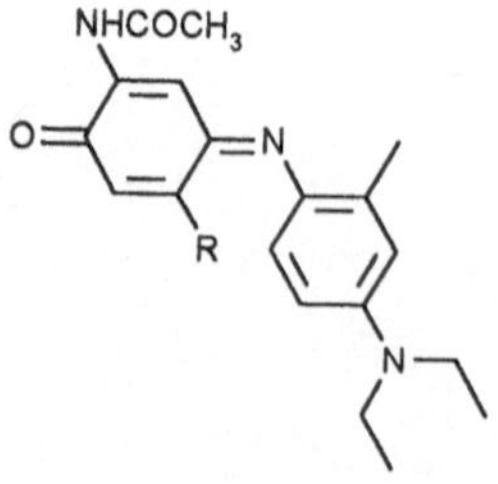

(**40**): R = H

(**41**): R = CH$_3$

Indoanilines derived from a variety of naphthols, e.g. (42) and variously substituted phenols, e.g. (43) are also claimed to exhibit good properties.

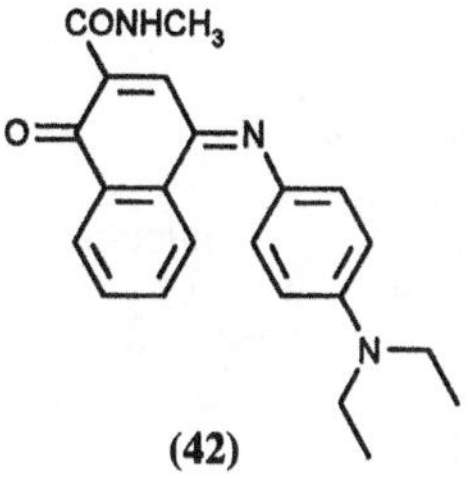

(42) 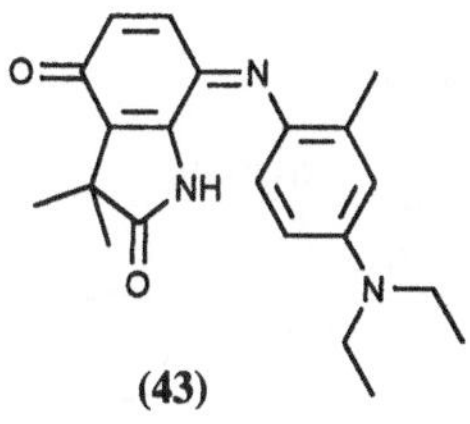   (43)

A number of anthraquinone dyes have been proposed as D2T2 cyan dyes. The 1,4-bis-(alkylamino)anthraquinones (44) are reddish blue in hue and as such are of limited utility as cyans. Greener hues are provided by 1-alkylamino-4-arylaminoanthraquinones (45) but these are still not true cyans. However, the introduction of electron-withdrawing groups into the anthraquinone nucleus leads to dyes of more bathochromic hue. This effect has been exploited with dyes of improved hue and lightfastness being obtained with structures such as (46).

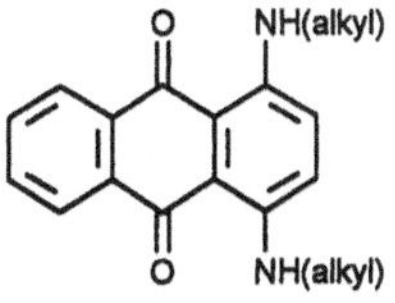

(44) 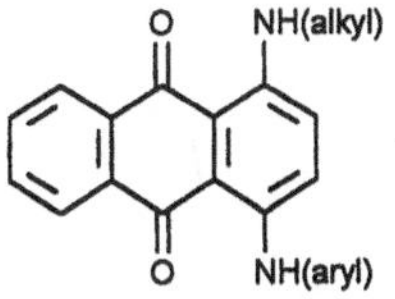   (45)

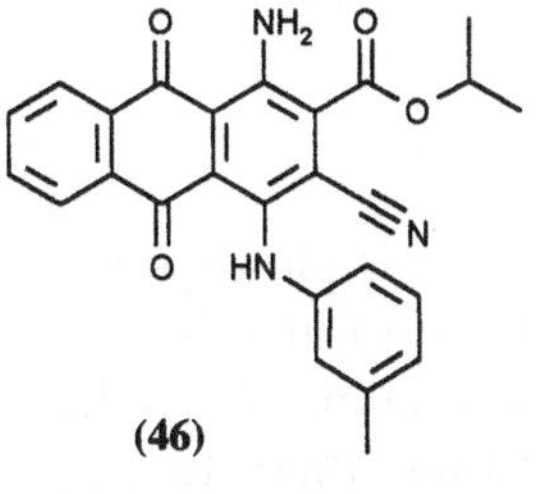

(46)

Although cyan dyes derived from 2-amino-3,5-dinitrothiophene, e.g. (47) give high print optical densities they exhibit poor lightfastness and better results are obtained with azothiophene dyes such as (48). The inherently strong disazothiophene cyan dye (49) gives stable prints of high optical density and lightfastness. Print stability may be improved still further by the incorporation of branched alkyl or ester groups, e.g. (50) and (51), and improvements in hue achieved by introducing electron-withdrawing groups into the diazo component.

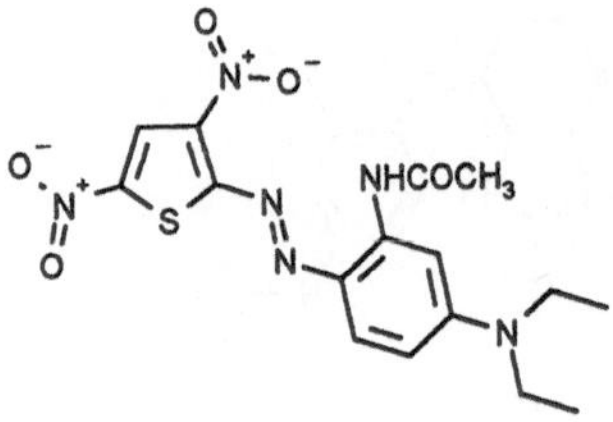

(47)

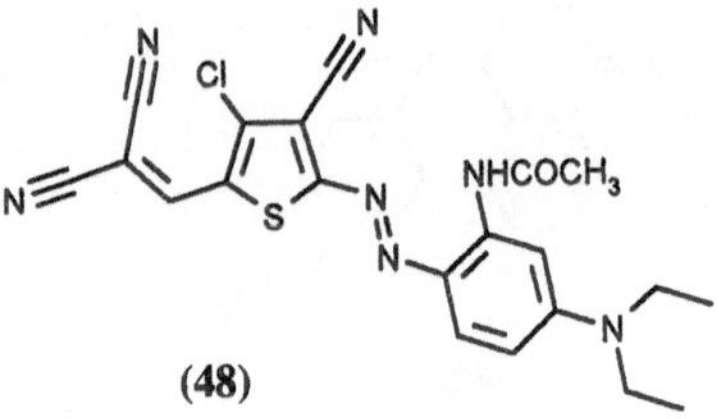

(48)

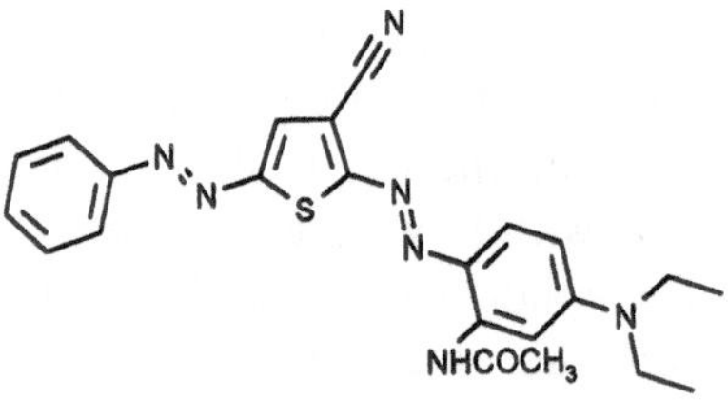

(49)

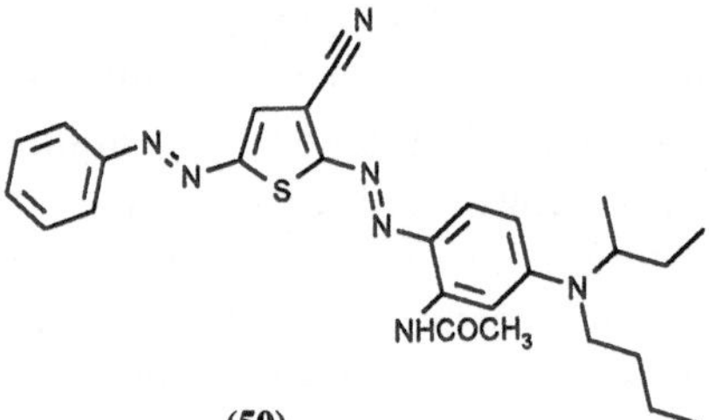

(50)

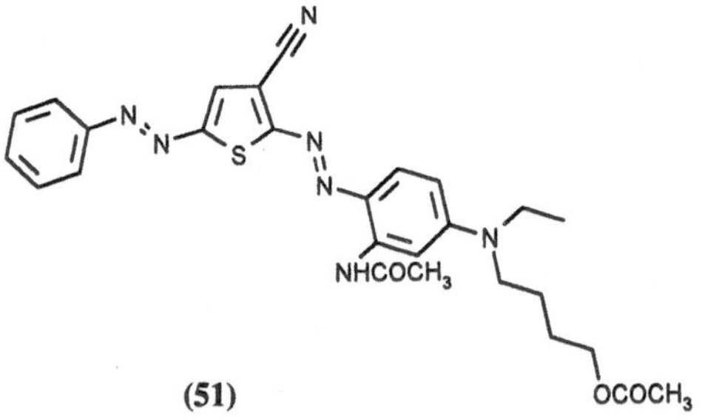

(51)

6.4.4 D2T2 market

Widespread use of electronic photography remains the jewel in the crown for D2T2 printing but unfortunately this has so far failed to materialise. The reasons for this are due mainly to the current state of development of electronic cameras rather than to any shortcoming of the printing technology. As a result, D2T2 has to date had no discernible impact on silver halide based photography. The quality of D2T2 images which can be achieved is, however, beyond doubt and once camera manufacturers develop solutions to existing problems, D2TD will be poised to take advantage. Until then, D2T2 will continue to find uses in other areas such as the novelty sectors (amusement parks, pictures on mugs, etc.), identity cards and credit cards and medical imaging. Colour proofing has also begun to utilise D2T2 printing technology, an example being the launch of the 'ColorEase' proofing system by Eastman Kodak. A large proportion of the D2T2 consumables, dyesheet cassette and receiver sheet combinations, originate in Japan with companies such as Dai Nippon Printing investing heavily in the technology. There is also evidence of considerable research activity by a number of companies including Fuji-Photo Film, Mitsubishi Kasei and Mitsui Toatsu Chemicals. In the USA Eastman Kodak have been very active as have ICI, Agfa-Gevaert and BASF in Europe.

6.5 Thermal-melt transfer

Thermal-melt transfer was the first of the two main thermal-transfer technologies to be developed. The process involves a colour ribbon having a coloured meltable coating either in the form of repeating sequential yellow, magenta and cyan panels for full-colour imaging or as single, usually black, panels for monochrome images. The ribbon also generally includes some form of registration mark to allow the printer to function accurately. Thus, on first inspection the ribbon is of similar form to that utilised in D2T2 printing. There are, however, fundamental differences in the constitution of the colour layer and in the mode of transfer of the colour which will now be described in more detail.

6.5.1 Dyesheet

The donor sheet consists of a thin (<25 μm) base film, usually polyester of 6 μm thickness, although other substrates such as condenser paper, polyethylene film, polypropylene film, cellophane and polyimide may be used. The main considerations are that the substrate should be readily available at low cost, be strong enough to be handled without tearing in the printer, have a smooth surface to facilitate release of the wax layer on printing and in particular have good thermal conductivity. In contrast to D2T2, a protective, thermally resistant backcoat is not necessary as the temperatures involved are lower than the softening point of polyester. The colour layer consists of a wax or other relatively low melting substance containing either a dye or pigment (cf. D2T2) or mixture of both. Examples of suitable waxes include carnauba wax, paraffin wax, montan wax, beeswax, ceresine wax and isocyanate-modified waxes. Paraffin wax has been widely used due to low cost and ready availability of paraffin waxes of varying melting points. Some of the paraffin waxes can cause a progressive deterioration in recording properties and to overcome this pigmented, chlorinated paraffin wax mixtures have been proposed (Nippon Denki KK Japanese Patent 58 162 568). The coloured wax coating must be reasonably hard, sufficient to prevent soiling, but have a low melt viscosity when heated to temperatures of around 70°C. This is generally achieved by blending a relatively hard wax with a softer wax and introducing additives such as metal powders to improve the thermal conductivity although these may have a detrimental effect on the transparency of the printed image. If the melting point is too low, the storage stability of the spooled donor sheet is reduced resulting in transfer of some of the wax film to the reverse side of the base film with which it is in contact. If it is too high, >100°C, then sensitivity is poor.

Unlike D2T2, specially coated papers are not required for printing on to (see section 6.5.3).

6.5.2 Melt-transfer colorants

Colorants for melt transfer have to satisfy a number of criteria: they should be thermally stable, be non-bleeding so as to avoid loss of resolution on printing, have good lightfastness, be non-toxic and, obviously, be compatible with the host wax. Since colorant transfer does not occur independently of the wax, then either pigments or dyes may be used. In general, pigments possess most of the desired properties and dominate the area. Typical examples of pigments for full-colour printing are the azo yellow (**52**), the quinacridone red (**53**), and the phthalocyanine blue (**54**). For monochrome output the most common pigment is carbon black.

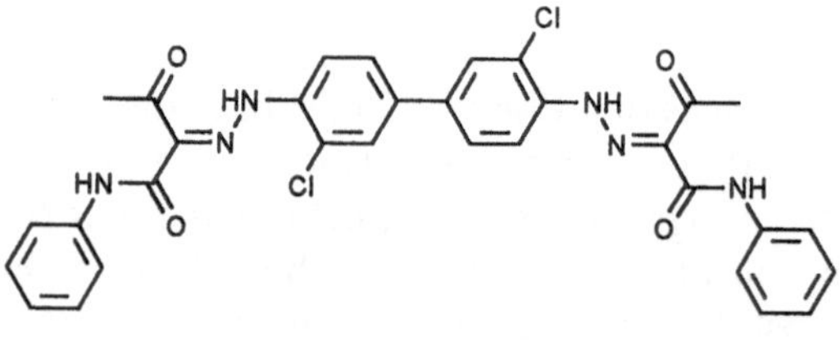

(**52**)

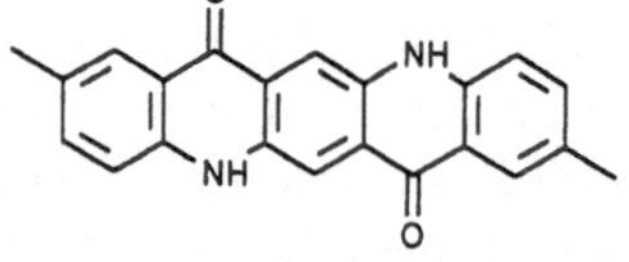

(**53**)

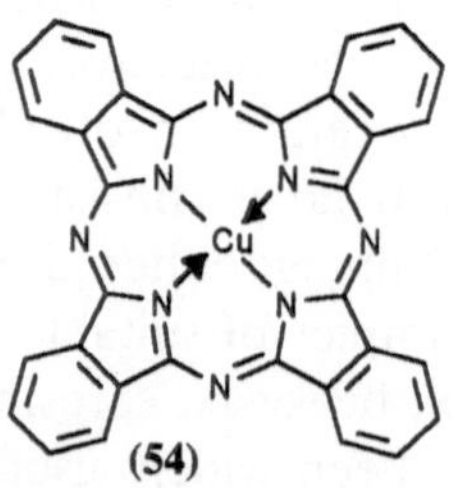

(**54**)

Whilst pigments exhibit high fastness properties their use in wax transfer suffers from one major limitation which is a lack of transparency. This limits the accuracy with which colour reproduction can be achieved and is particularly evident in prints made for viewing as overhead transparencies. This effect is very noticeable in the yellow area where what appears to be an acceptable bright-yellow on paper, i.e. viewed in reflectance, shows as a dull brown when viewed in transmission. An obvious way to overcome this problem is to substitute dyes for pigments which, provided they are soluble in the wax, give completely transparent images. The chlorinated waxes are particularly suitable for use with dyes as these are non-crystalline and thus highly transparent in their own right. Unfortunately, the greater mobility of the dye molecules within the wax host can lead to bleed problems or problems due to crystallisation on storage. One way in which the high transparency and strength provided by dyes has been combined with high fastness properties is to synthesise coloured condensation polymers which replace the conventional waxes (ICI European Patent 333 337). As the colour is an integral part of the transferred medium, bleeding is no longer a problem and the dyes are no longer capable of crystallising from the medium. The dyes for this system may be chosen from any of the known chemical classes of dyes provided that they contain at least one active hydrogen, e.g. hydroxy, primary or secondary amino, mercapto or carboxylic acid groups. By polymerising a lactone (e.g. caprolactone, 55 or hydroxyalkanoic acid in the presence of a dye containing such functionality, e.g. magenta, 56, or yellow, 57), a condensation polymer containing the dye as an integral part of the polymer chain is obtained. The amount of dye may be varied so as to achieve any desired depth of colour in the polymer. The resulting coloured condensation polymers are soluble in solvents such as tetrahydrofuran and can therefore by applied to the base film by, for example, gravure-coating procedures. Overhead transparencies made from these polymers have excellent clarity.

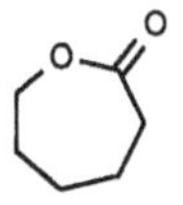

(55)

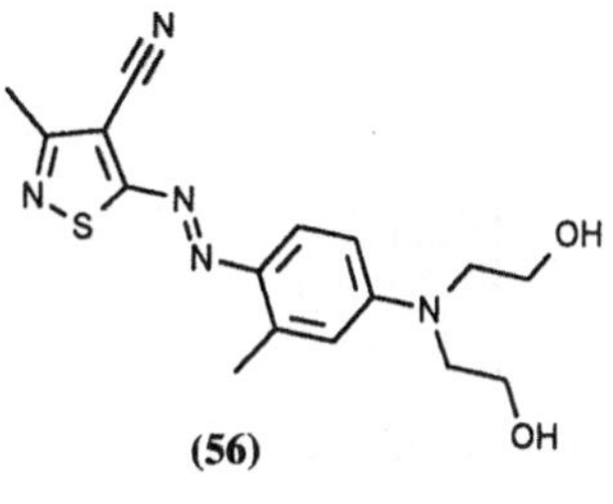

(56)

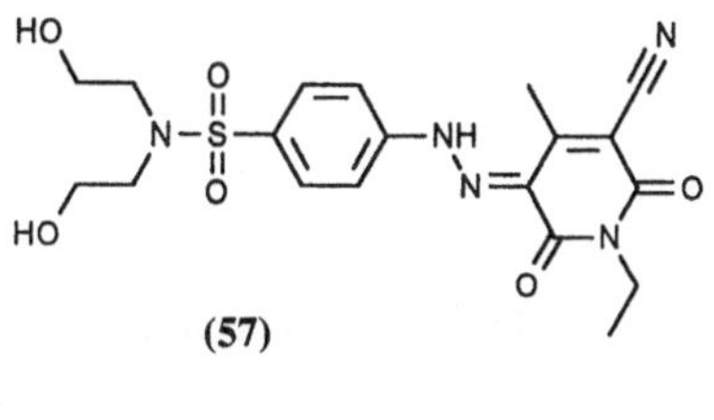

(57)

6.5.3 Receiver sheets

For thermal melt-transfer processes the receiver sheet is normally plain paper, or transparency in the case of overhead projection. Indeed, the ability to use plain paper is one of the major advantages of the process. In order to form the print, the colour layer must adhere preferentially to the receiver when being separated from the donor sheet. Much of this is due to the relatively rough surface of even 'smooth' paper where the molten colour can penetrate below the immediate surface and key into the paper. In the case of transparencies the adhesion is determined by the ability of the colour layer to wet the film surface.

6.5.4 Printing

As thermal melt transfer is an all-or-nothing transfer, the system cannot produce a true grey scale of coloration (*cf.* D2T2). In practice, a semblance of a grey scale is achieved by subdividing the gross pixel structure into smaller print areas (Figure 6.9).

By printing multiple smaller areas of colour, the eye is unable to detect the discrete dots and interprets this as a grey scale. Also, because the transferred layer resides essentially on the surface of the paper, the prints have a waxy feel and are prone to damage by abrasion. Nonetheless, thermal melt-transfer printing provides a relatively inexpensive means of producing either monochrome or full-colour hard copy.

6.5.5 Thermal melt-transfer market

Compared to direct thermal printing, the current market for thermal melt transfer is relatively small. However, this situation may change significantly

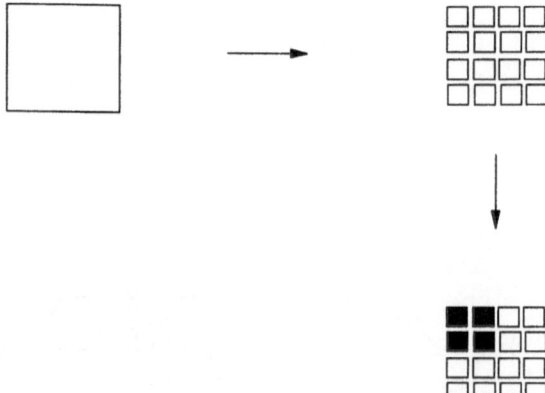

Figure 6.9 Subdivision of pixel area in thermal-melt transfer.

as new applications emerge, although much will depend on competition from alternative technologies such as ink jet. Most of the manufacturing capacity resides in Japan with Fujicopian Company dominating the market prior to 1983 when Dai Nippon Printing Company entered the field; several others have since followed. It is believed that, although there are potential manufacturers in the USA, activity is inhibited by a shortage of US suppliers of thin polyester base film. One of the principal difficulties facing potential competitors is the need to acquire the necessary coating technology.

References

Bradbury, R. (1995) Dyes for dye diffusion thermal transfer (D2T2) Printing. In *Modern Colorants: Synthesis and Structure*, H. Freeman and A.T. Peters (eds). Blackie, Glasgow.

Fox, B. (1988) *New Scientist*, 17 December 1988.

Hasimoto, K. (1988) *Shikizai*, **61**(4), 234.

Igarashi, A., Usami, T. and Ishige, S. (1994) *IS&T's Tenth International Congress on Advances in Non-Impact Printing Technologies*, p. 323.

Murata, Y., Maeda, S., Hirota, T. and Morishima, T. (1989) *Mitsubishi Kasei R&D Review*, **3**(2), 71.

Rastogi, A.K. and Wright, R.F. (1989) *SPIE* (1079), 183–214.

Takahashi, Y., Segawa, T., Shirai, A. and Toyofuku, K. (1994) *IS&T's Tenth International Congress on Advances in Non-Impact Printing Technologies*, 349–351.

Watanabe, K. (1993) *SPIE* (1912), 76–82.

7 Optical data-storage systems

P.A. HUNT

7.1 Introduction

Optical data storage (ODS) represents the leading edge in consumer-based electronic memory systems. The use of laser energy to write and read microscopic marks provides the most areal efficient data-storage technology commercially available to date. The spectrum of technical options which exist within the ODS system allows end users to write and archive data for over 30 years, or to write and erase data thousands of times: each option also offers transfer of data at millions of characters per second and at a price which is <1% of the price of storing data on paper or microform.

The fundamental principles of ODS were determined in the late 1960s with much of the development being carried out by Philips Research Laboratories. The late 1970s saw the emergence of small, cheap, relatively high-powered laser sources in the shape of the diode lasers, which provided a pathway to low-cost, miniaturised hardware, suitable for application in professional and consumer markets.

ODS technology has been developed into a wide range of formats. In disc format it became available for use in read-only-memory (ROM) applications in 1984. ROM discs are currently used as pre-recorded audio compact discs (CDs), video CSs and as CD-interactive discs in consumer applications, and as CD-ROM in personal computer (PC) and professional applications. The 1990s has seen the development of user-writable and user-erasable CD technology, which promises to broaden the possible applications in the professional imaging, PC and consumer markets.

The launch of the world's first flexible ODS product in the form of ICI 1012 Data Storage tape in 1991 saw a product capable of providing 1 terabyte (TB) of data storage per tape, giving significantly higher volumetric efficiency than magnetic-tape competitors in the archival data-storage market. ODS technology has been adopted in the production of smart cards. The cards have been used in a manner similar to credit cards and are used for storage of personal details, e.g. medical records. The greater data capacity available on an optical strip over that available on traditional magnetic systems provides storage of 2–3 megabytes (MB) per card, compared to significantly less than 1 MB on a magnetic card. ODS has been used in the development of tag products for use in the tracing and logging of equipment (a traditional bar-code reader application), and as a replacement for diazo microform in the storage of analogue images.

This chapter presents an overview of the techno-commercial situation within the ODS market. An outline of the basic principles of the technology is presented, together with a description of the physical and chemical features of the main system types. There is an analysis of the technical performance offered by ODS, and an assessment of the competitive position of the technology. Details of applications for the technology are described, with a vision of future ODS products and markets completing the chapter.

7.2 Principles of operation and functionality

7.2.1 Read-only memory (ROM)

The first ODS system to impact on consumer markets was audio read-only memory discs, more commonly called CDs, where information is pre-recorded and the user cannot add data. The product provided significant improvement over conventional audio systems in sound quality and durability, due to the recordings being digitally based and containing a high degree of error correction. The principle by which a CD functions uses 'pre-encoded' structural patterns to represent the digital data. The pre-encoded data is read with a low-power laser passing over the features, picking up a modulated reflected signal through variable light scattering as a result of interaction with the surface topography.

A vacuum-deposited aluminium or silver coating is present beneath the 'textured' surface of the disk to provide a light-reflective layer. The disc has a polycarbonate or polymethylmethacrylate substrate, into which grooves are set. A protective layer of radiation-cured acrylate completes the assembly.

A significant advantage of optical processes is that reading of the information is a non-contact process; this allows thousands of re-reads with no loss of signal quality due to surface wear effects, a common issue with gramophone or magnetic-tape recordings. Laser resolution allows the storage of data in a highly areal efficient manner. With currently available lasers, which operate at 780 nm, surface features of <1 μm in size are resolvable. A typical ROM disc is capable of storing c. 650 MB of data, which can be accessed at a data rate of >1.2 Mbits/s.

The early 1990s saw a number of reported problems with the lifetime of CD products. For instance, the Philips CD product was found to undergo degradation of the silver reflector layer caused by a reaction with breakdown products (sulphurous/nitrous) from within the polymeric protective layer. Although formulation changes were made to correct the degradation problem, the issue highlighted concerns over claims of limitless lifetime for CD products.

7.2.2 Erasable versus write-once

The range of technologies developed for ODS systems provides the means to store data for either write-once–read-many (WORM) or erasable use. In WORM systems, written data form permanent indelible marks, which can be read repeatedly over periods of many years. In erasable technology, the written mark can be erased through a secondary irradiation step, and new data written in its place. This is the case with magnetic recording systems. The production of erasable marks has been achieved through the use of metallic alloy systems capable of reversible morphological changes on irradiation. Photochromic dyes and liquid crystals have also proven suitable organic-based materials for erasable applications. These systems are described fully in section 7.3.

7.2.3 Physical and chemical requirements

In matching the performance of available ODS WORM or erasable active-layer technologies with the requirements for specific data-storage applications, there are a range of physical and chemical performance characteristics which require definition. In the production of a mark there are several basic needs; the most fundamental is that the material should absorb laser energy within the wavelength range of the laser chosen for the application. Furthermore, the sensitivity of the technology (i.e. amount of energy to make a mark) should be sufficient to provide the desired data rate (i.e. rate at which data is written) with the laser power available. Once the mark is formed the chemical/physical changes produced by the laser should provide sufficient contrast between written and unwritten areas that subsequent reading of the data can be achieved. A further area for consideration is in the storage and use of the media. It is of central importance that the stability of written and unwritten media should be consistent with the desired product lifetime. Stability in such cases can be defined as the ability to retain mark clarity during repeated reads, light or chemical robustness, or resistance to damage during repeated winding in tape products. Finally, the chemistry may be required to be within certain toxicological/environmental limits, which is of particular importance in consumer based markets.

7.3 System options

A wide variety of materials has been identified as being suitable for information storage in optical systems. These materials range from metallic-based systems, through to dye and liquid crystalline coatings. The key characteristics of WORM and erasable devices employing these materials are described below.

7.3.1 Metallic ablative (WORM)

The principle of mark production is through 'mass movement', i.e. ablation of a pit. Laser energy is absorbed in the metal layer, raising the temperature within the metal to above its melting point. Sensitivity of metallic ablative systems is dependent on the melting point of the metal, the absorption at the laser wavelength and the active layer thickness. Tellurium, on which much of the early work was based (Lou et al., 1981), has a melting point of 450°C and a strong absorption over the wavelength range 450–850 nm.

The technology later progressed through the use of tellurium alloys, in an attempt to extend the lifetime stability of the medium (Markvoort et al., 1983). Tellurium and its alloys are not highly archival, being susceptible to the formation of oxides by hydrolysis. The generally quoted product lifetime for such technology in a hermetically sealed disc is ten years. Tellurium alloys, and metallic technology in general, are highly light stable, and have low susceptibility to damage through repeated reading. The metallic nature of the technology provides a distinct threshold for mass movement: therefore, read laser power is simply set to ensure that surface heating during reading falls short of the threshold.

Examples of products which have utilised tellurium ablative technology are numerous. Philips have released a WORM disc product which uses a tellurium/selenium alloy and Daicel also use the same metals in their WORM product. Hitachi produce a 12-inch 7 gigabyte (GB) capacity disc (OC321–2), with an active layer of tellurium/selenium/lead alloy, based on chemistry developed in the early 1980s (Gotoh and Nakamichi, 1989). Mitsui Petrochemicals have released a product with a tellurium/carbon alloy (Lightstore AP-WO50).

A more recent development in metallic ablative systems has been the introduction of tin-based alloy technology. The two exponents of the system are Dow Chemicals and Dainippon Ink & Chemicals. The motivation for using tin alloys is their improved lifetime stability. Improved chemical stability allows the technology to be used in non-hermetically sealed environments (e.g. in tape products).

The Dow technology uses a tin–bismuth–copper alloy. Early reports of the technology suggested a metallic separation mechanism during mark formation (Strandjord et al., 1992). The mechanism produced high sensitivity, but at this stage no protective layer was used. A more recent paper outlined the introduction of inorganic (dielectric) protective and stabilising layers (Larson et al., 1993). These layers provide improved chemical stability for the active layer, but appear to reduce sensitivity. The original formulation (non-overcoated) had a sensitivity of < 1 nJ/mark, whilst the new formulation requires > 2 nJ/mark. The Dainippon Ink & Chemicals technology uses a tin–selenium alloy. The mechanism is a classic ablative process, which gives high contrast and sensitivity of < 0.8 nJ/mark.

Closely related to metallic ablative is the alloy forming WO (write-once) technology. The technology uses two adjacent metallic layers which when irradiated undergo an alloying process. The alloying has the effect of significantly changing the optical parameters of the two layers and thereby changing the reflectivity of the written mark. Sony have developed a product based on this technology called the Century disc (so called as it is claimed to last 100 years). The disc has adjacent antimony–selenium and bismuth–tellurium alloy layers.

7.3.2   Metallic phase-change (WORM and erasable)

Metallic phase-change technology uses metallic alloys which possess the capability to exist in both amorphous and crystalline forms at room temperature. The technology has been used in WORM systems using a single amorphous-to-crystalline transition, and in erasable systems where it is possible to carry out reversible amorphous-to-crystalline and crystalline-to-amorphous transitions. The mark contrast is produced through using metal alloys which produce optical parameters for each morphological form that are sufficiently different to provide significant reflectivity contrast. In general, sensitivity is higher where there is a low amorphous-to-crystalline transition temperature, and written data stability is higher where there is a high-energy barrier for the reverse transition.

Early work on phase-change metallic technology used tellurium-based alloys and was an extension to work being progressed on ablative tellurium alloys. The erasable capability of the technology provided the opportunity to match the erasability of magnetic media, which remained a key performance advantage for magnetic systems at the time. The first prototype erasable technology was developed in 1982, although the basis for the technology was developed in 1970 by Energy Conversion Devices in the USA (Feinleib, 1972). The late 1980s saw significant research in this area, with Sony, Matsushita, Toshiba and Hitachi being amongst the companies involved in the technical development.

Although tellurium-based phase-change formulations are highly patented, relatively few formulations have been developed into products. Matsushita have been most successful in applying the technology. They have released a WO-disc product based on a tellurium–palladium alloy (Nishino, 1986), plus an erasable product which uses a tellurium–antimony–germanium alloy. Asahi (Terada et al., 1993) and Hitachi also sell erasable products based on tellurium–antimony–germanium alloys. Matsushita and Asahi have licensed their formulations widely, but to date there is no single set of standards covering WORM/erasable phase-change technology.

More recently there has been a range of antimony-based technologies developed. Kodak have released a 14-inch WO disc product which uses an antimony–indium–tin alloy (Tyan *et al.*, 1992). The low-temperature amorphous-to-crystalline transition (150°C) and optical properties of the metals produces a high-sensitivity capability (< 0.5 nJ/mark). The chemical stability of both amorphous and crystalline phases provides extended lifetime capabilities. Philips have reported two WO phase-change alloys with high sensitivity (< 0.5 nJ/mark) and archivability (Gravesteijn, 1988). The systems are based on alloys of antimony–indium and antimony–gallium, but as yet they have not been adopted in a product. In a broadly similar way to ablative systems, the addition of a protective coating (e.g. silicon oxide or nitride layer) to phase-change media, has a desensitising effect. The protective layers form both a heat sink and act to suppress the slight volume increase associated with the production of molten alloy, a key stage in the mark-forming process.

Recent reports suggest that Matsushita and Philips are looking into the potential of using phase-change materials for use in CD-ROM. They are also being developed for use in CD-R products, where they have significant competition from the Sony MO (Magneto-optical) technology.

7.3.3 Dye–polymer ablative (WORM)

The development of dye–polymer ablative technologies followed closely behind metallic ablative, the pit forming process again depending upon a mass movement mechanism. The principle used is shown in Figure 7.1. The writing process uses a near infrared high-powered semiconductor laser (typically >10 mW) to form a shallow, 1 μm diameter pit in the surface of the dye coat. The absorption of the laser energy by the dye causes a temperature rise of >300°C at the surface, which prompts melting and flow of the dye–polymer layer. The pits formed constitute recorded data. The data is read through the use of a low-powered laser which scans the surface of the dye. The laser receives a modulated reflective signal from the dye layer surface as a result of the variable dye thicknesses present in regions exposed and unexposed to laser energy.

In organic materials the presence of a strong absorption band produces strong optical activity (i.e. a rise in refractive index) in the region of the absorption band. The relationship between absorption and refractive index is outlined in standard texts (Yariv, 1985), and is shown in Figure 7.2.

Early work on dye–polymer systems focused on dyes with absorption bands in the 488 nm range, for use with available solid-state lasers of the

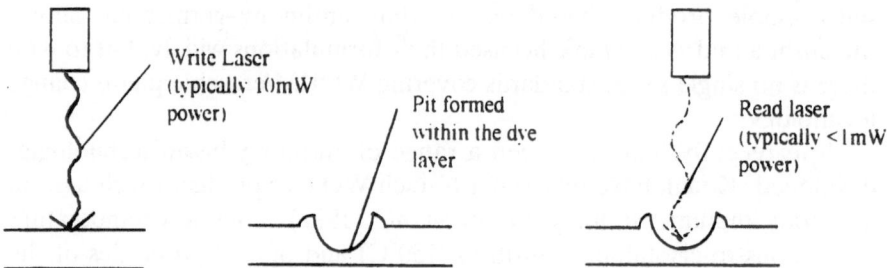

Figure 7.1 Principle of WORM optical-data storage in dye–polymer media.

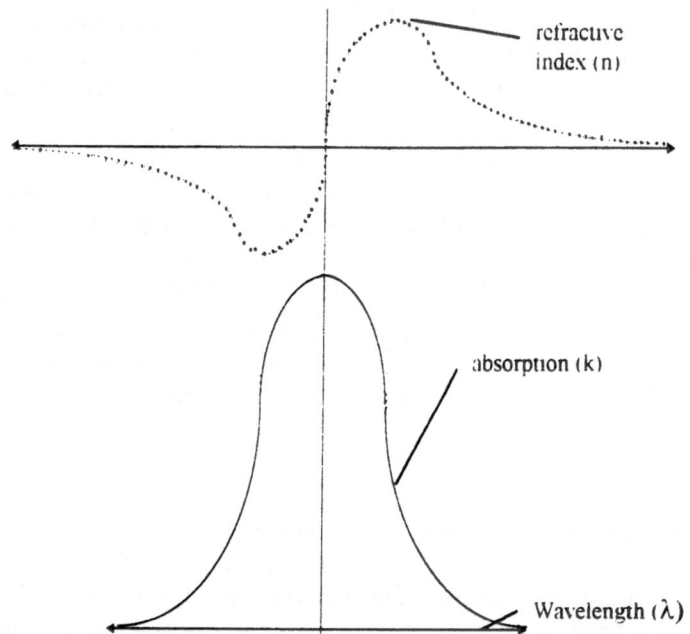

Figure 7.2 Relationship between absorption and refractive index for an organic dye system.

time. Many of the early application patents in the dye–polymer area were filed by Kodak, who used azo dyes in exemplification of their inventions (Thomas and Wrobel, 1983). The first experimentation using infrared absorbing dyes occurred in 1980 (Jipson and Jones, 1981).

Dye–polymer technology can be susceptible to repeated read damage, depending on the glass-transition temperature (T_g) of the dye–polymer mix. Read damage, i.e. low-temperature damage, is caused by a loss of kinetic control within the polymeric matrix, allowing disturbance of the pit definition. In products which are susceptible to read damage, the use of low read powers is required. However, the large reflectivity contrast produced by dye–polymer systems allows high-clarity read signals despite the low-power analysis. Much of the early work in the infrared range used squarilium chromophores (1). These dyes produce narrow, strong absorption bands in solution, which broaden significantly in the solid state. However, despite the broad absorption, the small molecular size and relatively low melting point of the dyes produced a high-sensitivity capability (Gravesteijn, 1988).

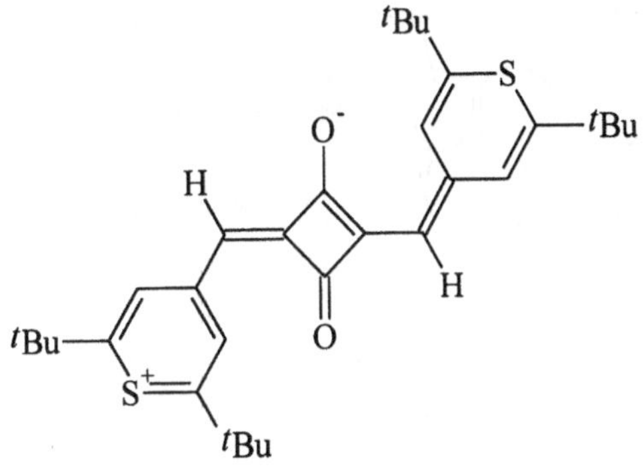

(1)

The later introduction of infrared diode lasers (in the range 780–830 nm) prompted extensive development of new infrared absorbing dyes. Much emphasis was placed on the innovative development of phthalocyanine and naphthalocyanine chromophores. Unsubstituted naphthalocyanine (2) and phthalocyanine (3) complexes are pigments, with broad absorption bands centred in the 650–690 nm range. The addition of sterically bulky pendant groups, such as siloxy, aryl or alkyl ether or thioether moieties, produced a bathochromic shift into the infrared, and developed increased solubility by reducing the ability of the macrocycles to form close molecular associations. The chemical changes carried out on the chromophores produced dyes capable of exhibiting extinction coefficients (E_{max}) > 200 000 and a bandwidth of < 50 nm in the case of phthalocyanines (Duggan et al., 1984), and similarly an E_{max} > 250 000 and a bandwidth of < 30 nm for naphthalocyanines (Kenney et al., 1988).

Phthalocyanine dyes have been adopted in a number of products. For example, ICI has utilised an 830 nm active phthalocyanine within their 1012 Optical Tape Product and Mitsui Toatsu in Japan have developed a large portfolio of high-performance dyes, the dye technology having been developed for use in interactive CD products amongst other quoted applications. In the late 1980s Hoechst Celanese developed naphthalocyanine dyes for use in disc-based products (Nickles et al., 1990). More recently, a number of companies including Hitachi and Mitsubishi Kasei have carried out extensive research into soluble naphthalocyanine derivatives. The chemical and light stability of macrocyclic dyes produces significant benefit in product archivability (Emmelius et al., 1989).

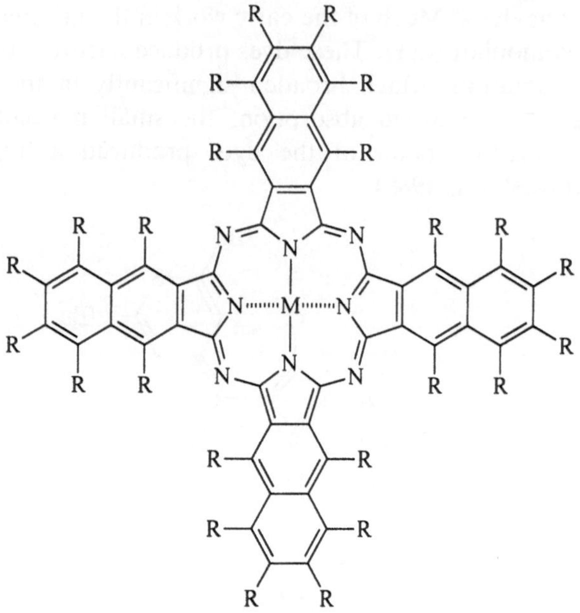

(2)

R = H, alkyl, carboxy   M = Ge, Si(OSi(R')₃)₂

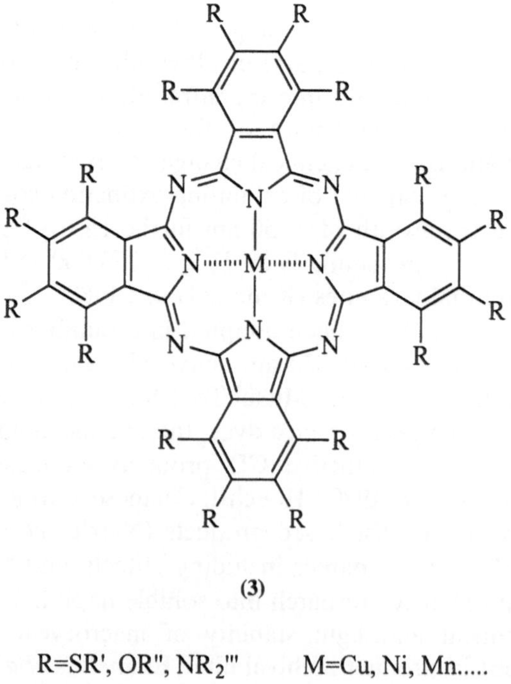

(3)

R=SR', OR", NR₂'"   M=Cu, Ni, Mn.....

Dyes within the cyanine group of chromophores (4) have recently been adopted as part of the new CD-recordable (CD-R) product developed by Taiyo-Yuden. Cyanine dyes were used due to their extremely narrow bandwidths (< 40 nm in the solid state) and low absorption at the 780 nm laser wavelength. The latter is required in order to provide the high-reflectivity coating required to maintain the CD standard. Interestingly, Taiyo-Yuden found the high contrast produced by the cyanine dye in the CD-R disc was caused by diffusion of the dye into the polycarbonate substrate during the writing phase (Hamada *et al.*, 1992). TDK have produced a WORM optical disc product (WOD-512) which uses a nickel dithiolate-stabilised cyanine dye. Pioneer (DC-502A product) and Maxtor Corporation (800MB. 5.25-inch optical disc) have released WO products using cyanine technology. A limitation in cyanine chemistry is poor light stability, which results in limitations in environments in which the disc products can be stored.

In the late 1980s, Ricoh adopted a methine dye (5) within a WORM disc product (RO5041) for professional applications. The methine dye, as with cyanine chromophores, suffers from poor light stability; the dye systems have the ability to form high-quality coatings from solvents and have the capability to produce high sensitivity.

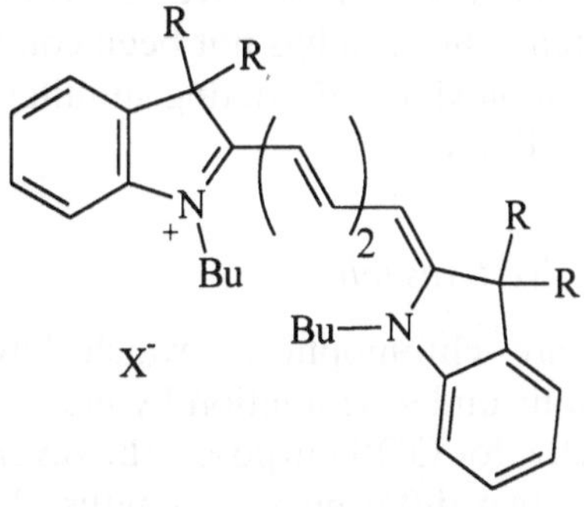

(4)

R=Me X⁻ = ClO_4^-

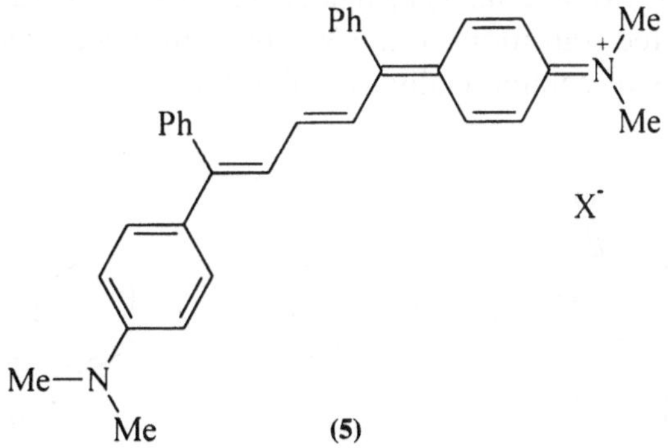

(5)

The addition of polymer to the dye is normally required to provide adhesion to the substrate on to which the dye is coated. The polymer also contributes to the optical performance, being critical in defining sensitivity and contrast. The flow requirements of pit formation result in a strong dependence on the T_g and melt viscosity of the binder–dye mixture (Molaire, 1988). Typical binders are thermoplastic polyesters, polymethyl-methacrylates and polystyrenes. In products where the dye–polymer layer undergoes contact from external surfaces, there is often a protective layer added to enhance abrasion resistance. An example of this is the ICI 1012 tape product, in which a radiation-cured acrylate layer is used to protect the dye–polymer layer. The addition of a top coat reduces sensitivity as was the case with metallic ablative systems.

An erasable dye–polymer technology has been developed by Optical Data Incorporated. It works on the principle of 'bump forming'. The bump is produced through the use of an expansion and a retension layer, each of which interacts with a laser of different wavelength. The 'expansion' layer (beneath the retension layer) undergoes laser heating during the writing process, inducing thermal expansion. This produces compressive stress within the 'retension' layer, and a tensional stress within the expansion layer. To erase, the retension layer is laser heated to above its T_g reducing the modulus of the layer, allowing the stresses within the system to relax and the bump to flatten. The idea has not been commercialised, probably due in part to the complexities of coating multilayer organic structures (Halter and Iwamoto, 1988).

7.3.4 Photochromic dye (erasable)

Photochromic dyes are chromophores which have the capability to reversibly change colour under irradiation by means of internal structural rearrangements. Ideally, for ODS purposes, the reversible colour change is triggered by lasers at two differing wavelengths. The erasable potential of the technology has prompted significant research into these chromophores, the most investigated to date being the fulgides (6), spiropyran (7) and azobenzene (8) systems (Feringa et al., 1993). These chromophores have light-initiated transitions in the 300–650 nm range, where as yet there are no examples of commercially available high-powered diode lasers.

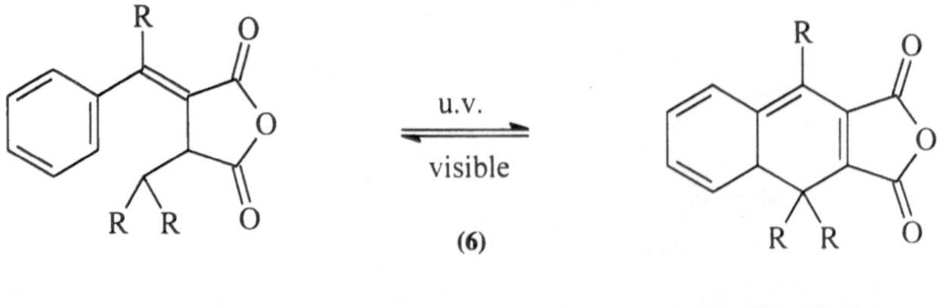

(6)

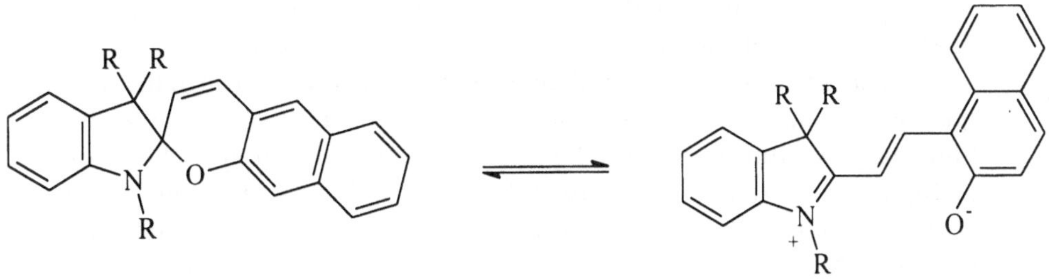

Colourless   (7)   Coloured

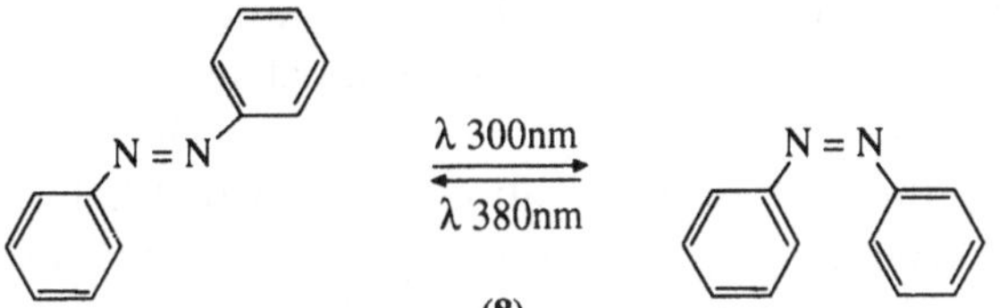

(8)

A major weakness in currently available photochromic mixtures is generally poor lightfastness. Attempts to improve performance in this area are being made through the production of photochromic activity within more stable moieties, e.g. phthalocyanines (Irie, 1988).

Among the companies which have announced formal commercial interest in the development of these systems are Sanyo, Hitachi and Matsushita. However, with no immediate promise of stability improvements, the application of this technology is likely to be restricted until the late 1990s.

7.3.5 Liquid crystals (WORM and erasable)

The principle used in applying liquid crystal (LC) technology in ODS systems is that laser heating imparts alignment variations within organic macromolecules, thereby producing a mark contrast. The principle is generally equivalent to an organic phase-change system. The recent growth in LC ODS development has been prompted by the belief that the technology is a low-toxicity route (when compared to metallic ODS) to high-sensitivity media (< 1 nJ/mark).

Products using LC technology can be characterised by the size of mark used in the data-storage process. Low-resolution technology, with a mark width of 5–20 μm, is being developed for applications such as microform replacement. Companies with a commercial interest in this area are GEC and Akzo. High-resolution technology, with a mark width < 5 μm, is being developed for applications such as CD-R, optical tape and smart cards.

GEC have patented high- and low-resolution systems which provide mark contrast by using the light-scattering properties of LC films. The film is manufactured with aligned polymeric chains, achieved through the use of specific heating regimes or by application of an electric field. The aligned film causes scattering of laser light when irradiated. On application of a high-powered write laser, there is total disruption of alignment, which gives a clear film and no scattering. The laser heating within the film is triggered by addition of infrared absorbing dyes, tuned to the laser wavelength.

Microform products remain important in 'deep-archiving' document applications, particularly in financial markets. In microform, the image is retained as an analogue copy, but which is significantly reduced in size by the image-production process. The traditional means of manufacturing microform used a diazonium dye active layer which underwent local photobleaching through the application of ultraviolet light to produce the desired image. The image production required photographic development (i.e. wet chemical processing). The LC product has advantages over diazo technology in that wet chemical development is avoided and the system offers low toxicity, which has always been an issue with diazo dye technology.

Traditional microform products have a lifetime capability of over one hundred years in controlled conditions. The long lifetime requirement has been emulated in the LC system by using stable chemistry, based on polysiloxane or polyacrylate LC polymers, which have high T_g and low susceptibility to hydrolysis. The coatings include dyes with low susceptibility to photodegradation. The LC microform is potentially erasable, and therefore reuseable. However with an erasure temperature >100°C, accidental erasure becomes unlikely (Bowry and Bonnet, 1991).

BASF developed high-resolution LC technology in the late 1980s. The system uses polyester or acrylate-based polymers and quinone-type dyes. Alignment is achieved through coating the LC polymer on to a prerubbed polyimide layer or through application of an electric field. The addition of a metallic reflector layer provides a reflective read–write system (in contrast to the transmission system used for low-resolution systems). The aligned polymer (unwritten state) undergoes a transition into an isotropic glassy state during the writing process. The reading of the data is achieved through assessment of contrast produced by either light polarisation or dye–polymer dichroism (Etzbach et al., 1985).

An alternative, high-resolution system has been developed by Akzo from which they plan to develop a CD-R product (Picken, 1993). The system is based on a twisted nematic film, which in the product would be placed between cross-polarisers. In the unwritten state, the LC polarisation allows light transmission, but once written the twisted structure is destroyed and no transmission is seen on reading.

There are numerous infrared absorbing dyes used within the LC systems, many of which are small, broad-band absorbing chromophores. Dyes in LC systems can be connected to the polymeric chain (grafted) or be chemically separate. The shape of the dye molecule is key in systems where dye alignment with the polymeric units is central to mark formation. A more recent development in the use of LC polymers is in the stabilisation of photochromic dyes (Natarajan *et al.*, 1992).

7.3.6 Metallic interference (WORM)

Plasmon Data Systems developed a unique method of producing a WO data storage media which uses an active layer consisting of a metallised interference grating pattern, which looks like the microscopic surface of a moth's eye. The laser marks are formed when the grating is irradiated. The scattering of the light within the surface relief produces surface heating, which melts the interference grating and allows a strong reflectance of the read laser light (Storey *et al.*, 1988).

7.4 Technical performance features of ODS systems

7.4.1 Performance advantages of ODS systems

The diverse range of optical data storage products already in the marketplace is proof of the massive potential of the technology. The differentiating feature of optical storage is the areal efficiency which the laser resolution provides. The areal capacity of ODS basically is limited by the wavelength of the radiation. A mark made with an 830 nm laser can have an area of < 1 μm^2, whilst using a 488 nm (blue) laser, the mark size can be < 0.4 μm^2.

Data marks produced by magnetic systems are not able to match ODS areal efficiency. In 2D (disc) storage applications this provides ODS products with significant product differentiation over magnetic competition. However, in 3D (tape) storage applications, volumetric efficiency is the key parameter, and both substrate thickness and area efficiency are of central importance. Magnetic-tape products are currently manufactured over a range of thicknesses from 25 to 14 μm, *cf.* ICI 1012 tape 72 μm substrate. The introduction of new magnetic head designs (e.g. helical scan) has increased areal efficiency of magnetic tapes, although they still remain inferior to ODS products. Table 7.1 shows a comparison of volumetric efficiency for ICI 1012 optical tape against those of both 2D optical and magnetic 'cartridge-based' products. The 1012 tape product is shown to provide the most efficient volumetric solution for data storage

Table 7.1 Comparison of volume efficiencies: volumes for 1 terabyte of storage

Media type	Unit capacity (MB)	Number of units/terabyte	Volume per terabyte (cubic feet)
3.5-in. Floppy disc	1.44	715 000	1240
3480 Magnetic tape	200	5000	32
5.25-in. Optical disc	650	1538	12
4-mm DAT tape	1350	741	1.86
8-mm Magnetic tape	2150	466	2
12-in. Optical disc	6550	153	9.4
ICI 1012 Optical tape	1 000 000	1	0.4

applications of > 1 TB. In 1994, ODS systems retained an areal advantage of 4:1 over the most advanced magnetic products (e.g. 8 mm Exabyte).

A key parameter with large on-line databases is the information response time. The comparison of data retrieval time from a single optical and a magnetic 'Winchester' disc shows that the magnetic system has a slight advantage. However, in large databases (up to 1 TB), the massive on-line data capacity of a juke-box of 14-inch optical discs, or optical tape, provides comparatively rapid access times to any record within the data when compared to automated silos of magnetic tapes.

The extensive range of ODS active layer technologies available has provided the opportunity to combine high read–write performance with long-term data stability. An example of this is the Kodak WO phase-change metallic system. If offers over thirty years' archival potential and a data rate of > 1 MB/s. In contrast, magnetic media can offer a data rate of up to 120 MB/s data rate, but the technology continues to suffer from lifetime stability problems and wear damage from repeated reading.

A further advantage in ODS technology is the availablility of WO technology. In security-based applications (e.g. bank records, government), potential erasability can be undesirable. In such cases, data indelibility has been used as a feature to differentiate WO-ODS from erasable (magnetic) competition.

There is a market-perceived advantage for ODS products over magnetic competition in the toxicological and environmental impact of using the technologies. ODS has undoubtedly benefitted from adverse publicity covering the personal and environmental toxicity of chromium-based magnetic media. In comparison, materials used in optical data storage are in general of low toxicity. More recent toxicological concern over the use of metals such as tellurium and selenium in ODS systems has reduced the ODS advantage.

7.4.2   Performance limitations in ODS systems

A major limitation in current ODS products is their data rate. Data rate is limited by a number of factors, amongst which are media sensitivity, number of write lasers, and laser pulse frequency. A number of technical innovations have been introduced which should stimulate higher data rates. For example, the wider adoption of non-mass-movement technology provides higher sensitivity media (< 0.5 nJ/mark capabilities). Hardware is emerging which uses laser arrays, (e.g. the Creo 1003 Optical Tape Drive, which uses 33 write lasers) and novel high-frequency laser pulsing devices such as the Philips polygon system, a highly efficient surface-scanning technique (Van Rosmalen *et al.*, 1994).

There is a current limitation in the number of commercially available erasable ODS products. Erasability continues to be a key requirement for consumer and PC data storage products, and remains a strong performance advantage for magnetic systems in these markets. The importance of erasability is further exemplified in the professional data storage market, where there has been increased sales of the Matsushita rewritable phase-change disc product, despite its relatively high cost and low data rate. There are currently no erasable ODS tape products available on the market, and a limited number of technologies suitable for the application. The application of high-sensitivity metallic phase-change technology or emerging systems such as liquid crystals/photochromics may be key in the commercial release of such products over the next five years.

7.5   Commercial performance features

An estimation of the ODS market size, with associated major players and products is given in Table 7.2. There is considerable market growth in all optical product groups. The majority of CD products are undergoing an annual growth rate of $>30\%$. The more novel optical products such as tape and card products are undergoing a slower growth profile but the opportunity for new products remains high in these areas, which would lead to greater potential for accelerated growth.

7.6   Comparison with alternative technologies

The number of products competing with optical data-storage media in both professional and consumer data-storage applications is large. However, they generally emerged from a small number of technical variations. Product differentiation in such markets is achieved through subtle changes in product format, broad comparisons of key performance criteria and

Table 7.2 ODS market overview

Product	Markets	Cost/unit[1]	Data capacity/unit	Main producers	Estimated sales value, 1994 (market/year)
Video/audio-CD/CD-I	Consumer	£5–30 To be announced	600 MB 3 GB	Philips, Sony, Kodak	>$B1
CD-ROM	Professional	£5–100	600 MB	Philips, Sony	>$B1
Photo-CD	Consumer	£5	650 MB	Kodak	>$M10
CD-R	Consumer/professional	£15	680 MB	Taiyo-Yuden, 3M, Kodak	>$M10
Optical tape	Professional	£6000	1000 GB (1 TB)	ICI[2]	<$M1
Optical disc	Professional/PC	£100 (5.25 in.) £180 (5.25 in.) £400 (12 in.)	WO, 0.65–0.94 GB Erasable, 1.3 GB WO, 5–10.2 GB	Matsushita	>$M100[3]
Magneto-optical disc	Professional Consumer/PC	£170 (5.25 in.) £23 (2.5 in.) £28 (3.5 in.)	500–650 MB 128 MB	Sony, Canon, 3M, Sony	>$M100[3] >$M10
Emerging/novel products	Professional	£5–100/card	2–6 MB	British Telecom, Canon, Drexler	>$M1

1. Pricing details gained from product-trade literature or from contact with product resellers.
2. ICI is the only company currently marketing a tape product. However, several companies within the USA and Japan have formally reported activities in the development of tape drives and consumables.
3. In terms of market share of the total optical-disc market (MO and optical), the Japanese companies dominate. In early 1992, it was estimated that Sony supplied 32% of the market. Matsushita 15% and the largest Western company, IBM, 3%.

through focused marketing. There are a number of parameters used by the data-storage industry to compare products. Commonly used ones include:

- System cost (drive and media) ($/MB)
- Media cost ($/MB)
- Volumetric efficiency (m^3/GB)
- Access time (s)

- Technical compatability
- Data rate (MB/s)
- Data lifetime (years)
- Maintenance cost ($/year)

The main competitive technologies to ODS are magnetic and magneto-optical. An example of a comparison between ODS products, conventional and magnetic systems is shown in Table 7.3, where price per terabyte of storage is listed for a number of common products.

7.6.1 Magnetic media

This technology was developed over forty years ago, and has been adopted in a range of formats. The first format was tape, which, through the design of small magnetic-tape drives, saw the introduction of media into consumer markets in the 1970s. The subsequent development of disc- and card-format products provided a route into every segment of the data-storage market.

Magnetic media uses an active layer of paramagnetic material, coated as a mixture with a polymeric binder (e.g. polyurethane). An electrically derived magnetic field is brought in close contact with the surface (0.1 μm spacing), and is modulated such that pulses of magnetism are transferred to these magnetic coatings. To erase the marks, the magnet is passed over the same area, with a different pattern of modulation. The reading of the information is the reciprocal of the writing process; a ring core (wire coil) is brought into contact with the magnetic surface such that the magnetic field from the marked data induces a modulated voltage in the core.

Table 7.3 Cost per terabyte of storage media

Media type	Unit capacity (MB)	Cost* per unit ($)	Cost per terabyte ($)
A4 Paper	0.0022	0.01	4 545 000
Microfilm	5.5	5.5	1 000 000
Floppy disc	1.44	0.93	654 000
3480 Magnetic tape	200	5	25 000
4-mm DAT tape	1350	12	9 000
8-mm Magnetic tape	2150	15	5 000
5.25 in. Optical disc	650	165	25 400
12 in. Optical disc	6550	360	55 000
1012 Optical tape	1 000 000	9000	9 000

*Pricing details gained from product/trade literature or from contact with product resellers.

The active layer has historically consisted of oxides of metals such as chromium, iron and cobalt. The shape of the oxide particles is known to be critical to high performance, as is purity. A number of products (e.g. Sony 8 mm video and DAT products) now use pure metal particles to produce larger magnetic signals than oxides. The mark formation is a non-mass-moving mechanism, allowing high sensitivity and data-transfer rates. The most advanced magnetic drives have data rates of > 120 MB/s, an advantage of c. 30 times over the fastest optical product available to date.

During the 1980s, the magnetic media industry suffered a number of major problems. For example, magnetic tapes used in archiving critical information, such as space flight data, were found to have degraded to beyond the point at which data could be retrieved. The hydrolytic degradation of the polyurethane binder, which produced adhesion of the magnetic coating to the reverse of the tape, was found to be the cause of the degradation.

The magnetic media industry has produced further innovation to conserve a degree of product differentiation over optical data-storage products. Helical scanning, a system of rotating magnetic heads, has provided a route to increase the areal capacity and data rate of the media. However, in achieving greater performance, the tape is being used at higher winding speed and with greater potential for damaging media–head contact.

Some improvements in archival performance have been achieved through cartridge design and chemical formulation changes. However, there remains concern over the coating stability under extreme conditions. The introduction of ferrite films is an example of an active layer which has greater resistance to hydrolytic degradation and increased data-packing capability.

Magnetic tape drives come in two design types, these being drives which use a linear writing (serpentine) system, and those that use the helical scan system. The most modern versions of linear drives use 3480/3490 tapes made by BASF and 3M and DLT tapes made by DEC for professional applications, and DCC tapes made by Philips and Matsushita for the consumer audio market. The most modern helical scan products are the DAT tape system, for which Sony is a major manufacturer, and the professional products, DCRSi (D1 format) and DST (D2 format) tapes made by Ampex, and the DIR1000 (DI format) supplied by Sony. Magnetic discs are made by a number of suppliers, the largest Western supplier being 3M.

7.6.2   Magneto-optical (MO) media

This technology has become a useful addition to the range of erasable optical-disc products available in professional data-storage applications.

MO technology offers higher data rates than optical discs, whilst maintaining competitive archival properties.

The principle of operation uses a laser to heat a 1 μm diameter spot on the surface of a paramagnetic alloy, such that the coercivity of the magnetic material becomes almost zero. Coercivity can be thought of as being the amount of energy required to change the magnetic field alignment of the magnetic layer. Whilst in the heated 'low coercivity' state, a bias magnet is brought close to the surface of the paramagnetic layer in order to switch the magnetic moment alignment within the spot (see Figure 7.3). The spot is then allowed to cool, freezing in the 'switched' magnetic mark. The written bit at this stage has a very high coercivity, and is stable to further external magnetic fields. To erase the written mark, it is again heated to a point where a reversed bias magnet can return the spot to its original alignment. The marks can be read by using a low-power laser, as a result of the application of the Kerr Effect. The different magnetic alignments over the written region interact with the polarisation of the laser light, producing a subtle rotation in light polarisation.

The chemistry of the active layer has seen considerable development. Current products use chemically unstable alloys of rare earth/transition metals. The alloys are stabilised by layers of dielectric materials, plus the full coating assembly is sealed hermetically. The MO discs are reported to be able to archive data for up to twenty years in the sealed system. The use of ferrite films in MO systems should present significant improvement in archivability, such films being capable of storage in a non-sealed environment. Ferrite technology also gives higher track densities, due to the ferrite magnetic particles being able to orientate in a direction perpendicular to the media surface.

There is developing competition between MO exponents, Sony particularly, and rigid optical disc manufacturers in the delivery of the most broadly accepted erasable disc based system. MO has recently been introduced into the consumer market by Sony, as the CD-R 'audio minidisk' product.

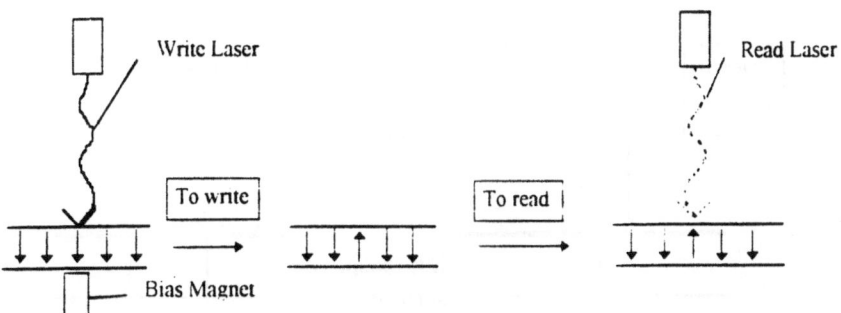

Figure 7.3 Description of the magneto-optical data-storage principle.

7.7 Related technologies

ODS has certain central features which are considered highly beneficial when broadly applied to other product areas. Amongst the transferable features are the high resolution of the laser source, laser interaction with infrared absorbing materials and the opportunity to generate computer-controlled digital outputs. Two areas where there is synergy between ODS technology and emerging imaging applications are lithographic plate manufacture and laser thermal-transfer printing.

Historically, lithographic plates were manufactured through wet chemical processing (see Chapter 2). The cost and complexity of the manufacture, and the modern trend in the printing industry towards shorter, more varied printwork has facilitated the introduction of flexible dry-processing routes to plate production. The new ODS-related technology uses infrared absorbing dyes and laser imaging in computer-controlled production of the plates. The infrared absorbing dye is added to a coating covering the printing-plate surface, the printing image being created through an infrared laser 'ablating' the coating, exposing ink receptive areas (Figure 7.4). The laser-induced 'thermal' system of producing plates has advantages in the fast-throughput, low-cost and high-quality capabilities of the technology.

The use of laser resolution has been adopted in the electronic imaging field. The introduction of the so-called L2T2(TM) technology (laser-light–thermal-transfer) uses infrared absorbing dyes (e.g. phthalocyanines) in the production of colour electronic images of near photographic quality (see Chapter 6). The use of infrared lasers has allowed the production of high-resolution images (up to 4000 dots per inch), with full-tone control (Hutt *et al.*, 1992).

The infrared absorbing dye is added to coatings of each of the three base-coloured dye panels (yellow, magenta and cyan) which make up a dye

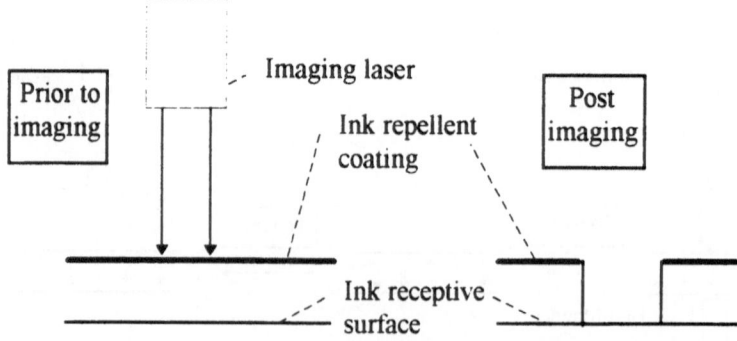

Figure 7.4 Principle of using laser ablation in the production of lithographic printing plates.

ribbon (film reel). The ribbon is brought into close contact with the imaging substrate, such that on application of laser pulses to the ribbon, heating of the infrared absorbing layer occurs, and transfer of the base colours to the substrate is achieved (Figure 7.5). The computer control of the process allows individual, accurate deposition of dye from the three panels of the dye ribbon in making a reproduction of the digital image. The recent introduction of high-powered diode lasers has provided the opportunity to increase the rate at which such images are generated. This technology has significant potential in applications such as the production of instant slides and in credit card manufacture, where personalised images on cards is a route to improving user security.

7.8 Applications of ODS

Optical disc technology has been widely adopted in both consumer and professional markets. The consumer-based CD products are used to store images, text and audio information. The traditional audio and video products have recently been joined by CD-recordable discs which allow the user to produce their own digital recordings. Kodak have released a Photo-CD product, which allows the consumer to store digital images in photographic form, with the added benefit of providing the user with the potential to display the image on visual display units.

CD products have assumed a central role in the creation of the multimedia concept for the consumer. The basis for multimedia systems is in the common set of standards developed for CD drive and media formats within all audiovisual applications, e.g. video, audio and photo-CD and CD-ROM. The standardisation allows the use of the full media range on the same basic hardware.

CD-ROM systems were introduced as professional educational aids in 1986. The discs are used to exhibit images, sounds and text. The lower cost of CD-ROM drives has prompted the introduction of CD-ROMs into consumer markets as an addition to a PC or as interactive video-game products.

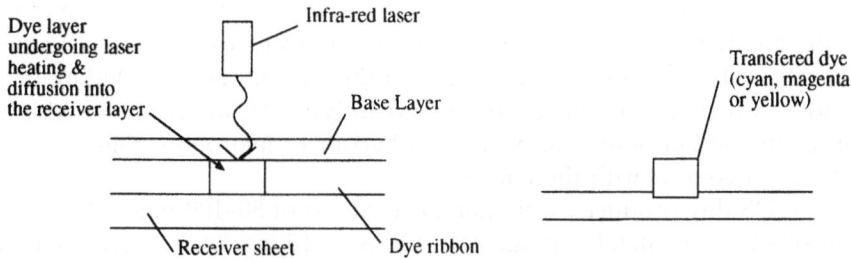

Figure 7.5 Principle of laser-light thermal transfer.

In professional applications, 5.25-inch optical discs have seen success in the magnetic media dominated PC market, as hard-disc backup and as image data buffers. The WO disc products are commonly adopted for long-term data archiving. Larger optical disc products (12–14 inch) have been introduced into juke-box drives, allowing the on-line storage of up to 1 TB of data. The juke-box systems have been adopted by a number of financial and government organisations, where the robustness of the drive and media, combined with the short access time, are key in archival solutions.

The ICI 1012 optical tape product offers highly competitive cost and performance advantages in applications requiring long-term storage of vast quanitites of data (Ruddick, 1992). A number of the largest archival storage applications in existence, with data libraries of up to 200 TB in size, have adopted optical tape. Amongst the users of optical tape are remote-sensing organisations (satellite data), medical imaging systems (where there is a requirement to store digital X-ray images) and in government financial record applications where long-term data security is essential.

Optical cards are becoming broadly used for patient medical record storage. The cards can hold over 1500 pages of text or up to 80 medical images, thereby making full medical histories instantly available to physicians.

7.9 Toxicology and environmental considerations

Toxicological considerations fall into two categories, namely toxicity with respect to those who manufacture the media and toxicity with respect to the final users. Product marketing has centred the public debate on toxicity in relation to the effect on the consumer. However, marketing has not distinguished between user and manufacturer with respect to the environmental effect of the making and disposal of products, where there has been open debate around long-term effects of metal and solvent emissions.

The toxicological facts with respect to 'end-user toxicity' are often overshadowed by a number of misconceptions. For instance, although metals such as antimony, tellurium and selenium have reputations for being highly toxic, the pure metals are of relatively low toxicity. The high toxicity of the metals manifests itself in their compounds, including their oxides, which are produced during hydrolysis. However, in disc-based products the hermetic seal provides a barrier to hydrolysis, and acts as a barrier to contact with the end user.

In ODS disc products, with metallic coatings of 50–100 nm in thickness, the quantity of metal present per disc is < 10 mg. Any oxide content, formed as a degradation product, would only be a further small fraction of this amount, and hence through ordinary use, the end user has an extremely low risk of exposure. The future use of metallic alloys in tape products may cause greater concern due to the larger quantities of materials involved and the lack of sealed edges on a tape. The risk of consumer exposure would be minimised by the use of cartridge and cassette packaging, whilst the choice of active layer would take into consideration major toxicological issues.

The majority of organic materials used in ODS systems are of low toxicity due to their low reactivity and large molecular size. Amongst materials classed within this category would be long-chain polymers, such as liquid crystalline polymers, and pigment-like dyes such as phthalo-cyanines. Protective coatings and substrates are normally fully reacted polymers and inorganics. In most cases, radiation-cured acrylates or silicon dioxide are used as protective coatings, whilst polycarbonate, polyester or polymethylmethacrylate thermoplastics are used as substrates.

The chemicals and processes used in the manufacture of ODS products make an interesting comparison with those used in manufacturing magnetic media. The preparation of organic ODS and magnetic active layer coatings use broadly similar organic solvents. The coatings are also applied using the same wet-coating techniques (gravure, web, etc.). However, magnetic products require a topical lubricant to minimise friction, and the majority of current products use fluorinated lubricants which can only be coated from chlorofluorocarbons (CFCs). The fluorin-ated lubricants are also practically indestructible, providing concerns about their long-term effect on the environment. Small quantities of CFC solvents are used in the preparation of rigid optical discs. A number of organic dyes require fluorinated solvents for spin-coating.

Metallic and dielectric layers in ODS systems have no associated solvent use as the layers are prepared by vacuum coating (sputtering or electron beam). Such processes allow control of external release of metals to the environment, but still provide toxicological concerns with respect to staff exposure during such routines as cleandown of the coating chambers. Strict industrial hygiene needs to be observed to reduce the risk of exposure during such procedures.

7.10 The future

As in many emerging technologies, initial sales growth was restricted by mistrust of the new technology. There was also general unease with laser technology. However, in the mid-1990s the use of lasers has become commonplace, a fact which owes much to the emergence of diode lasers. There is presently significant emphasis being placed on the development of low-wavelength diode lasers. The major focus of research is on the production of 415 nm (blue) lasers which have the capability to increase areal capacity by up to a factor of four over that produced by infrared lasers.

The CD industry has announced its intention to release a high-density CD product in 1995–96. The product will use a red laser (650 nm) to increase the capacity of the disc. The driving force behind the change is the desire to place a full-length digital movie (105 minutes) on a single video-CD. The industry has also stated an intention to adopt a blue laser source when they emerge as a low-cost product towards the end of the decade (Fox, 1994). The recent introduction of optical disc 'multimedia' technology appears certain to dominate the consumer information, games and audio/video markets into the late 1990s. The introduction of CD-R and erasable-CD products will remove any remaining advantage that magnetic media has held in these markets.

Photochromic dyes and polymers are currently undergoing considerable research, probably in part due to the push towards low-wavelength laser diodes. Laser sources within the 415–490 nm wavelength range would complement the photochemical requirements of a number of the chromophores. However, there remain serious concerns over the poor stability of this group of compounds and there is much work to do before commercialisation of the technology.

Tape-based products have dominated large data-storage solutions for nearly half a century. The volumetric efficiency exhibited by a tape format will always be superior to that of a disc. The combination of ODS technology and a tape format offers a significant volumetric advantage over current magnetic options. The current optical tape product, ICI 1012 tape, promises to be the first of an expanding range of tape products in which further increases in volumetric capacity are generated through the use of a media thickness closer to those used by magnetic media, i.e. 12–24 μm.

The introduction of small format cartridge-based optical-tape products will provide further reduction in the cost per megabyte value (potentially down to 0.1 cent/MB). The introduction of small-format optical tapes will enable the introduction of robotic tape-handling drives. Automated loading is seen as key in professional large data storage, an advance which, in the case of optical-disc products, has already provided the market with a route to a 1-TB on-line database.

There is a general requirement for both optical disc- and tape-based products to provide higher data rates in order to match magnetic media performance. Such improvements in data rate will be achieved through the use of high-powered, low-wavelength diode lasers, in conjunction with emerging optics technology.

The introduction of cassette-based optical-tape products into consumer markets seems a likely development in the late 1990s. A number of companies such as the Japanese Broadcasting Company and Sony have announced interest in developing a tape system for use in HDTV videotape products.

The household acceptance of the technology in the form of the CD and the broad potential for the ODS principle suggests that ODS systems are likely to be a major influence on data-storage products into the early 21st century.

References

Bowry, C. and Bonnett, P. (1991) Liquid crystal polymer optical memories: analogue, digital and holographic. *Optical Computing and Processing*, **1**(1), 13–21.

Duggan, P.J. *et al.* (1984) Optical recording medium. European Patent 186 404.

Emmelius, M., Pawlowski, G and Vollmann, H.W. (1989) Materials for optical data storage. *Angewandte Chemie, International Edition (English)*. **28**, 1445–1471.

Etzbach, K-H. *et al.* (1985) Optical recording medium. European Patent 17 1045.

Feinleib, J. (1972) Information recording system employing amorphous materials. US Patent 3 636 526.

Feringa, B.L., Jager, W.F. and de Lange, B. (1993) Organic materials for reversible optical data storage. *Tetrahedron*, **49** (37), 8267–8310.

Fox B. (1994) CDs: the next generation. *New Scientist*, 10 September, 33–35.

Gotoh, A. and Nakamichi, S. (1989) Long life 12-inch W-O type optical disks. *SPIE Optical Data Storage Topical Meeting*, **1078**, 36–42.

Gravesteijn, D.J. (1988) Materials developments for write-once and erasable phase-change optical recording. *Applied Optics*, **27**(4), 736–738.

Halter, J.M. and N.E. Iwamoto (1988) Thermal-mechanical modeling of a reversible dye–polymer media. *SPIE Optical Storage Technology and Applications*, **899**, 201–210.

Hamada, E., Fujii, T., Takagishi, Y. and Ishiguro, T. (1992) Recording process of recordable compact disc. *SPIE Optical Data Storage*, **1663**, 443–446.

Hutt, K.W, *et al.* (1992) Laser dye transfer. *Proceedings of IS&T's Eighth International Conference in Non-impact Printing Technologies*, Williamsburg, VA. Oct. 25–30. p. 367.

Irie, M. (1988) Phthalocyanine compound and optical recording material using it. European Patent 325 742.

Jipson, V.B. and Jones, C.R. (1981) Infrared dyes for optical storage. *Journal of Vacuum Science Technology*, **18**(1), 105–109.

Kenney, M.E. *et al.* (1988) Recording information media comprising chromophores. US Patent 4 725 525.

Larson, T.L., Woodward, F.E. and Pace, S.J. (1993) A new tape for optical data storage. *Japanese Journal of Applied Physics*, **32**, 5461–5462.

Lou, D.Y., Blom, G.M. and Kenney, G.C. (1981) Bit oriented optical storage with thin tellurium films. *Journal of Vacuum Science Technology*, **18**(1), 78–86.

Markvoort, J.A., Spruijt, A.M.J. and Vromans, P.H.G.M. (1983) Aging properties of optical non-erasable disks. *SPIE Optical Storage Media*, **420**, 134–140.

Molaire, M.F. (1988) Influence of melt viscosity on the writing sensitivity of organic dye–binder optical disk recording media. *Applied Optics*, **27**(4), 743–746.

Natarajan, L.V. *et al.* (1992) Liquid crystalline siloxanes containing spiropyran chromophores as reversible optical storage materials. *Advanced Materials for Optics and Electronics*, **1**, 293–297.

Nickles, D.E. *et al.* (1990) Naphthalocyanine chromophores for WORM-type optical data storage media. *SPIE Storage and Retrieval Systems and Applications*, **1248**, 65–73.

Nishino, S. (1986) Optical recording carrier and method of producing the same. *European Patent* 213 358.

Picken, S.J. (1993) Twisted nematic film, a method for the preparation thereof, and a display device comprising said film. *European Patent* 565 182.

Ruddick, A.J. (1992) ICI optical data storage tape – An archival mass storage media. *NASA Conference Publication*, **3198**, Vol. I. 265–274.

Storey, P. Longman, R.J. and Davies, N.A. (1988) Environmental evaluation of rugged and long-life write once optical disks. *SPIE Optical Storage Technology and Applications*, **899**, 226–232.

Strandjord, A.J.G. *et al.* (1992) Flexible storage medium for write-once optical tape. *NASA Conference Publication*, **3198**. Vol. I, 275–284.

Terada, M. *et al.* (1993) Optimized disk structure and Ge–Te–Sb composition for overwritable phase change compact disk. *Japanese Journal of Applied Physics*, **32**, 5219–5222.

Thomas, H.T. and Wrobel, J.J. (1983) Element for recording by thermal deformation. *US Patent* 4 380 769.

Tyan, Y-S, *et al.* (1992) Kodak phase-change media for optical tape applications. *NASA Conference Publication*, **3198**, Vol. II, 499–511.

Yariv, A. (1985) *Optical Electronics*, (3rd edn), Holt, Rinehart and Winston, London. pp. 135–139,

Van Rosmalen, G.E. *et al.* (1994) A compact optical tape recording system. *SPIE Optical Data Storage Technology*.

8 Electrostatic, ionographic, magnetographic and embryonic printing technologies
P. GREGORY

8.1 Introduction

Electrographic, ionographic and magnetographic printing are three non-impact printing technologies. They are less prevalent in the marketplace than the electrophotographic, ink jet and thermal technologies. However, they possess some advantages and have found niche markets in the printing industry.

Some embryonic technologies are a hybrid of two non-impact printing technologies. For example, Xeroprinting toner fusion is a hybrid of electrophotographic and electrostatic printing, whilst the VerdeFilm technology is a hybrid of electrophotography and thermal printing. Likewise, TonerJet is a hybrid of laser printing and ink jet.

8.2 Electrographic printing

In electrographic (also called electrostatic) printing a latent electrostatic image is created on a dielectric (insulating) surface, usually special paper or film, and the image made visible by applying toner (Gregory, 1991). The process is very simple and comprises just three steps:

1. Creating the electrostatic image. This is achieved by using a writing head consisting of an array of styli. Applying a voltage to selected styli produces ions which are deposited on to the dielectric paper or film.
2. The latent electrostatic image is developed with a liquid toner of opposite charge to the latent image. The small size of the particles in liquid toners enables high resolutions to be obtained (up to 400 dots per inch).
3. The toner is then fused to the paper or film.

The process is illustrated in Figure 8.1.

Electrostatic printing is especially good for large-format images, either monochrome or colour. Large-format printing is becoming increasingly digitized and this is creating additional markets and user bases for wide-format products. Electrostatic printers, such as 3M's Scotchprint Electronic

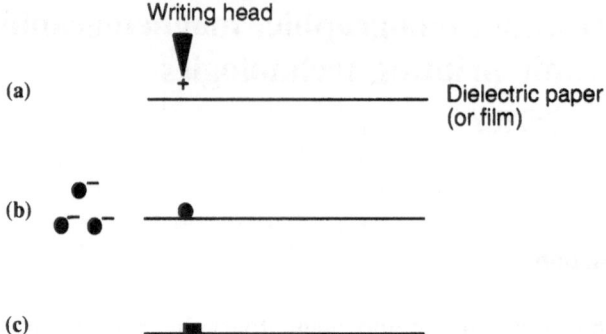

Figure 8.1 Electrostatic printing process: (a) write charged image, (b) develop with toner, (c) fuse.

Graphic System, overcome problems associated with conventional screen-printing processes (Skinner, 1993). A colour electrostatic printer for textiles has also been developed (Ishii *et al.*, 1993). A recent review (Anon, 1994) of the wide-format printing market compares the performance of electrostatic printers with other printing technologies and provides figures for the shipment and placements for the US market as well as a forecast for future development, market shares and sales. However, advances in other non-impact printing technologies, notably ink jet, is providing increasing competition in the wide format marketplace (see Chapter 5).

Electrostatic monochrome printing is the simplest and the fastest electrographic method, since it requires just a single pass. Full-colour printing is more complex since four colours are required to produce a full-colour image, namely yellow, magenta, cyan and black. This coloured image may be achieved by using a single toning station and repeating the process four times. Registration of the paper prior to each pass is important to ensure a high-quality image. This is difficult since paper is not dimensionally stable and therefore the transport mechanism must compensate for shrinkage and stretching of the paper. Nonetheless, Versatec uses the multipass system in their electrostatic plotters (Potter, 1987). Figure 8.2 shows a typical electrostatic plotter.

Alternative approaches are also employed. Thus, Benson and Synergy use a single pass but with four toning stations rather than one. This approach has fewer registration problems but is slower, since the paper plus toner must be dry before the next toner is deposited. Precision Image Corporation use a helical scanning technique to overcome the registration problem. This technique involves holding the dielectric paper or film on to a drum by vacuum and writing the electrostatic image with a helical scanning head. A single toning cup follows the scanning head and deposits

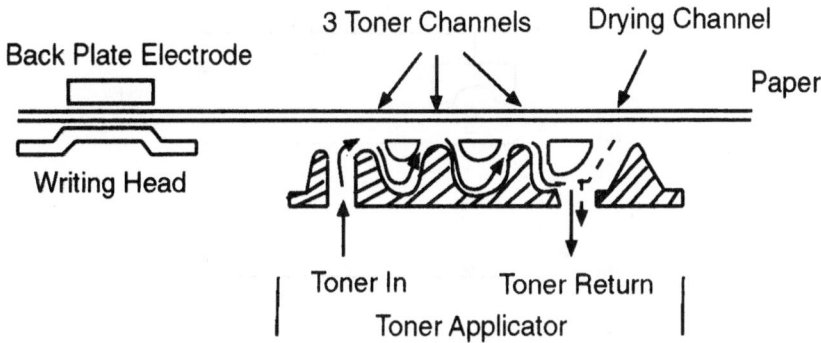

Figure 8.2 Electrographic printer (based on Versatec V80).

liquid toner. A separate reservoir contains cleaning fluid for flushing the toning cup between the colour passes.

8.3 Ionographic printing

Ionographic printing, also known as ion deposition, has been known for more than twenty years. It was developed by Dennison and has been pioneered and commercialised by Delphax (Gregory, 1991). It was first used for printing bar codes in the mid-1970s. The printing of bar-code labels is still important but other applications of ion-deposition printing include tickets, bills and cheques (Stobbe, 1993). Recent work on colour ion-deposition printing by NTT has resulted in a prototype colour facsimile machine (Omodani *et al.*, 1993). This prototype machine can send a full-colour (16.7 million colours) image in one minute. It is printed in high quality, including grey scales, using an ion-flow printing process which involves only a small number of simple process steps (Kubelik, 1990; Lo *et al.*, 1993).

Ion-deposition printing uses a simple four-step process. Step 1 consists of generating an electrostatic image on the surface of a dielectric drum. In step 2, the image is developed on that surface by applying toner to make it visible. Step 3 is the transfix step in which the toner is transferred and fixed to paper. Finally, in step 4, the surfaces are cleaned ready for the next cycle (Figure 8.3). Figure 8.4 shows the actual set up in a typical ion deposition printer.

The first step, the generation of the image, is accomplished using a print cartridge which deposits a pattern of electrical charge on the dielectric drum surface. It was thought that this charge was created from ions (Campbell Scott, 1988). However, it has been established (Caley *et al.*, 1991) that the latent image is produced by electrons, not ions. Because of this, ion deposition is also known as electron-beam printing (Smith, 1991).

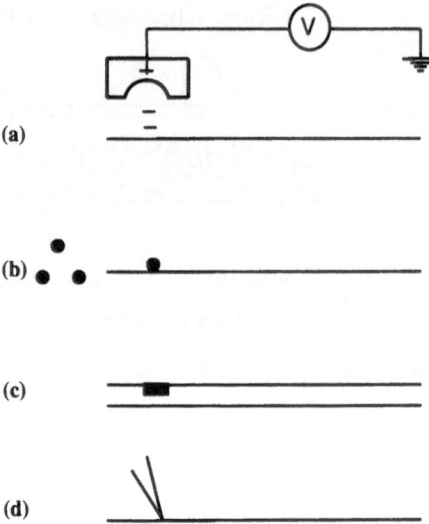

Figure 8.3 Ion-deposition printing process: (a) write, (b) develop, (c) transfix, (d) clean.

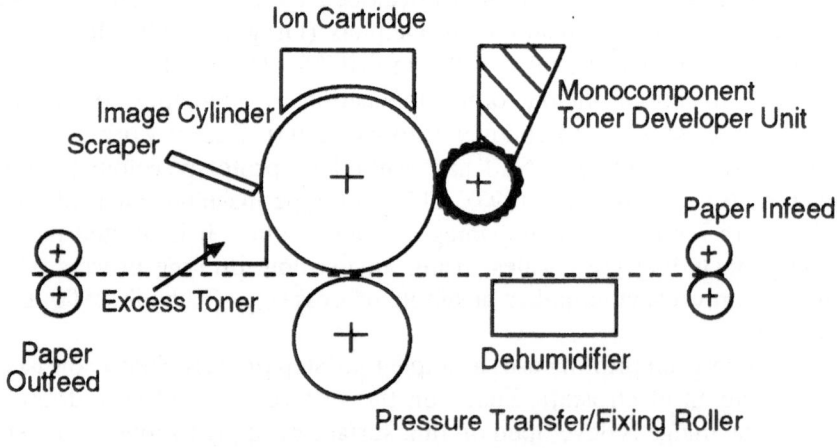

Figure 8.4 Ion-deposition type (Delphax) non-impact printer.

The dielectric drum is an aluminum cylinder with an extremely hard and smooth anodized outer surface which is specifically treated to hold the deposited electrons in position. It can be handled without fear of damage, unlike photoconductor drums. The hardness of the drum is significant because cold pressure rather than heat is used to fix the toner, usually a single-component toner, on to the paper. Greater than 99% toner fixation on to the paper is achieved, and hence little cleaning is required and the toner usage is low. For comparison, the toner transfer efficiency in electrophotography is 70–80%. Good capabilities exist with ion deposition both for grey scale and colour.

Ion-deposition printers are very rugged, fast (120 pages per minute, ppm), and reliable. Delphax and Denison guarantee the print cartridge for 125 000 pages and the drum for 1.25 million copies. This figure is in striking contrast with the 10 000 copies per lifetime of an organic photoconductor drum.

The main area for improvement is toner fixation to the paper. Cold pressure, though it uses less energy, is not as efficient as heat fusing the toner on to paper. Additionally, it imparts a sheen to the toner and paper which is undesirable in some applications.

The quality of ion deposition is not yet equal to that of laser printing. However, the simpler process and durability of ion-deposition printers is resulting in them finding niche markets.

8.4 Magnetographic printing

Like laser, electrographic and ion-deposition printing, magnetographic printing (magnetography) is also a toner-based technology. However, in contrast to these technologies, magnetography utilizes magnetic rather than electostatic forces to produce an image.

In magnetography, a magnetizable alloy (e.g. cobalt alloy) or metal oxide (e.g. gamma-Fe_2O_3) dispersed in an organic binder is used as the imaging surface. This is selectively magnetized by a magnetic writing head, similar to that in a tape recorder, to produce the latent magnetic image on the drum. As in the electrophotographic and electrostatic processes, the image is rendered visible by development with a toner, in this case a magnetic toner. The toner is transferred from the magnetic drum to paper using a corona transfer. Finally, the toner is fixed using heat and/or pressure. Figure 8.5 shows a typical magnetographic printer.

The advantages of magnetography include multiple prints from a single image and potentially very high printing speeds, for example, 500–800 ppm. Disadvantages include limited resolution (due to the design of the magnetic writing head) and the tendency of the drum to corrode, especially in humid environments, although Bull claims a drum lifetime of 10 million copies (Campbell Scott, 1988).

The French manufacturers Bull and Nipson are the main advocates of magnetographic printing. Bull are continuing efforts to further improve the speed and productivity of magnetographic printing and are exploring new potential applications. For example, Bull are working on the design, construction and testing of a full-size, fully functional prototype of a Press

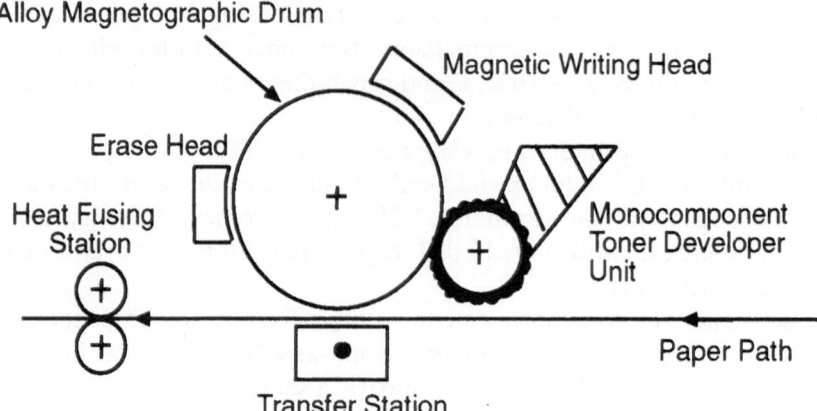

Figure 8.5 Magnetographic non-impact printer.

Integrable Magnetographic (PIM) module having the following character-istics (Eltgen, 1990):

- Consistent print quality at variable speed over a full range of speeds.
- Accommodation of high web tensions as may be required by a press.
- Printing process totally slaved to the paper web motion as imposed by press environment.
- Extended range of printing media.

Nipson offer a range of magnetographic printers capable of reaching production speeds of 700 ppm based on Bull patents. The print cylinder is covered by a protective layer 0.8 μm thick with a hardness of 2000 Vickers and a long lifespan. Nipson also produce the six types of pigment used, the 10 μm particles contributing to the quality of the print. The MP range of presses handle continuous fanfold printing while the Varypress range caters for reel-to-reel printing. Future enhancements include improved definition, higher speeds, two-sided printing and, in the long term, colour printing. The Varypress M700 is generally installed in-line with an existing press but can operate independently. It is PC controlled and runs at up to 105 m/min (Durchon, 1993).

8.5　Embryonic printing technologies

8.5.1　High-resolution dry-processing film (VerdeFilm)

A high-resolution dry-processing film that could render obsolete silver halide film and chemical processing for many pre-press graphic arts applications, and provide an environmentally friendly alternative for the printing industry, has been developed by Xerox Graphic Systems. Called

VerdeFilm, this material makes recording of colour separations or monochrome images a simple one-step process. This revolutionary product is claimed to be capable of resolution equal to, or better than, that of top-of-the-line silver halide products. It requires no special handling or storage and is daylight stable which greatly simplifies production and inventory requirements.

VerdeFilm uses the photoconductor technology of electrophotography, plus heat. As shown in Figure 8.6, the film structure consists of a polyester base, coated with a semi-transparent conducting layer, followed by a softenable thermoplastic coating and a protective overcoat. Near the surface of the softenable thermoplastic layer is a highly uniform monolayer of submicron photosensitive pigment particles.

Two types of VerdeFilm have been developed for graphic arts applications: scanner film, sensitive to blue–green light and imagesetter film, sensitive to visible and near infrared light. In the case of imagesetter film, the protective overcoat also contains an infrared-sensitive photo-pigment. Otherwise, the two films are virtually identical.

Unlike conventional film intermediates, VerdeFilm is not light sensitive except in the presence of an electrical field. Prior to sensitizing, VerdeFilm may be handled in room light eliminating darkroom requirements. Further, there is no need for light-tight packaging.

VerdeFilm is most conveniently used with an imagesetter or scanner that has been made Verde-capable. This requires adding a charging device and a heat-development system in the film path. In the case of imagesetter film, a flood lamp and an additional charging device are required.

As shown in Figure 8.7, scanner film is sensitized by depositing a negative charge on the film surface. A countercharge appears in the

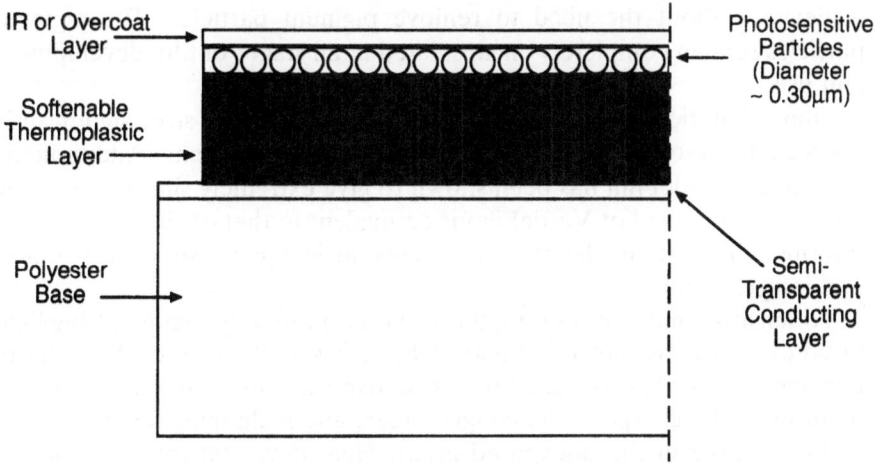

Figure 8.6 Film structure of VerdeFilm.

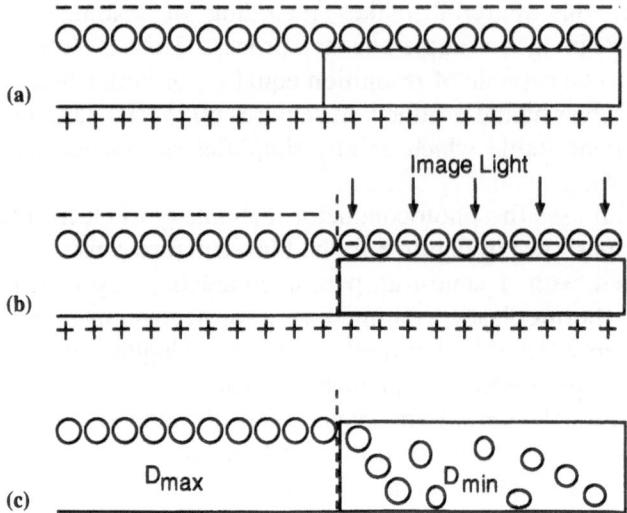

Figure 8.7 Verde scanner film: film process steps: (a) sensitize (charge), (b) expose, (c) heat develop.

semitransparent conducting layer. Imagewise exposure charges the sub-micron photopigment particles. Pigment particles not imaged remain uncharged (Ettinger, 1994).

During heat development, the softenable thermoplastic layer allows the charged particles to migrate in the direction of the positive counter charge, terminating in a random distribution. The unexposed particles do not redistribute since there is no electric force driving them.

Optical contrast is due to the difference in light transmission of the two layers. The undisturbed monolayer blocks light, and the randomly redistributed particle layer transmits light. Sufficient optical contrast is achieved without the need to remove pigment particles. The pigment particles remain completely encapsulated at all stages of film development and use.

Film resolution, dependent on the pigment particle size, significantly exceeds the resolution of current imaging/exposure systems. Additionally, developed VerdeFilm has been shown to give extremely sharp edges. The photographic speed of VerdeFilm is equivalent to that of silver halide film intermediate and enables the use of current imaging systems in scanners and imagesetters.

To sensitize imagesetter film, the surface is positively charged. Blue light flood exposure, as shown in Figure 8.8, causes all the pigment particles to become negatively charged. Imagewise exposure to visible infrared light neutralizes the charge in the imaged areas, and recharging neutralizes the surface charge in the unexposed areas. Heat development of the image setter film causes the unexposed particles to redistribute.

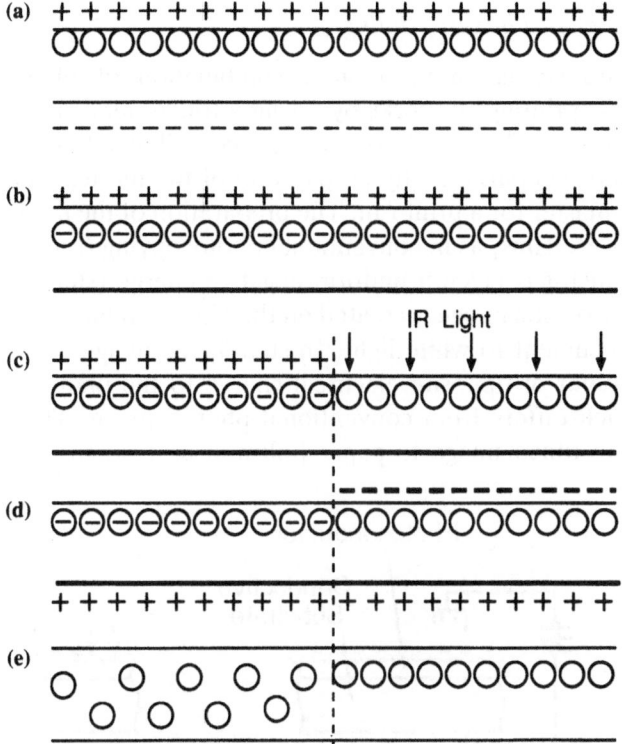

Figure 8.8 Imagesetter film: film proces steps: (a) sensitize, (b) blue light flood exposure, (c) IR image-wise exposure, (d) recharge, (e) heat develop.

The claimed advantages of VerdeFilm over conventional silver halide film are threefold:

- Ease of use.
- Environmental.
- Cost.

VerdeFilm combines the high-quality output of silver halide film with ease of use, namely its stability to light and dry processing. VerdeFilm's key ecological benefit is the total elimination of chemical processing which, in addition to producing waste chemicals, consumes large amounts of energy and water. Although VerdeFilm can be safely disposed of after use, Xerox plan to recycle exposed films and extract all the selenium for use in new film production. Finally, VerdeFilm should be more cost-effective than silver halide technology although the savings are described as significant rather than overwhelming (Coats and MacClean, 1994).

Although VerdeFilm is initially targeted at the commercial pre-press graphic arts market, it could eventually be applied to photography, although, as far as is known, no company is working on this aspect.

8.5.2   Toner fusion xeroprinting

This technology is based upon a combination of photocopying and electrostatic printing. It works by fusing a toned image to a conventional photoconductor (Bugner, 1991). The process, illustrated in Figure 8.9, is divided into two parts, firstly preparation of the master and secondly the use of the master for a print run. The preparation of the master is shown in Figure 8.9(a). Thus, as in conventional photocopying (see Chapter 4), the photoconductor is given a uniform negative charge (step 1). In step 2, a latent electrostatic image is created on the photoconductor by exposing the original document to white light. In step 3, the image is made visible by applying toner of opposite electrical charge to the photoconductor. It is step 4 which differs from conventional photocopying. Thus, rather than transfer the toned image to paper followed by heat fusion to produce a

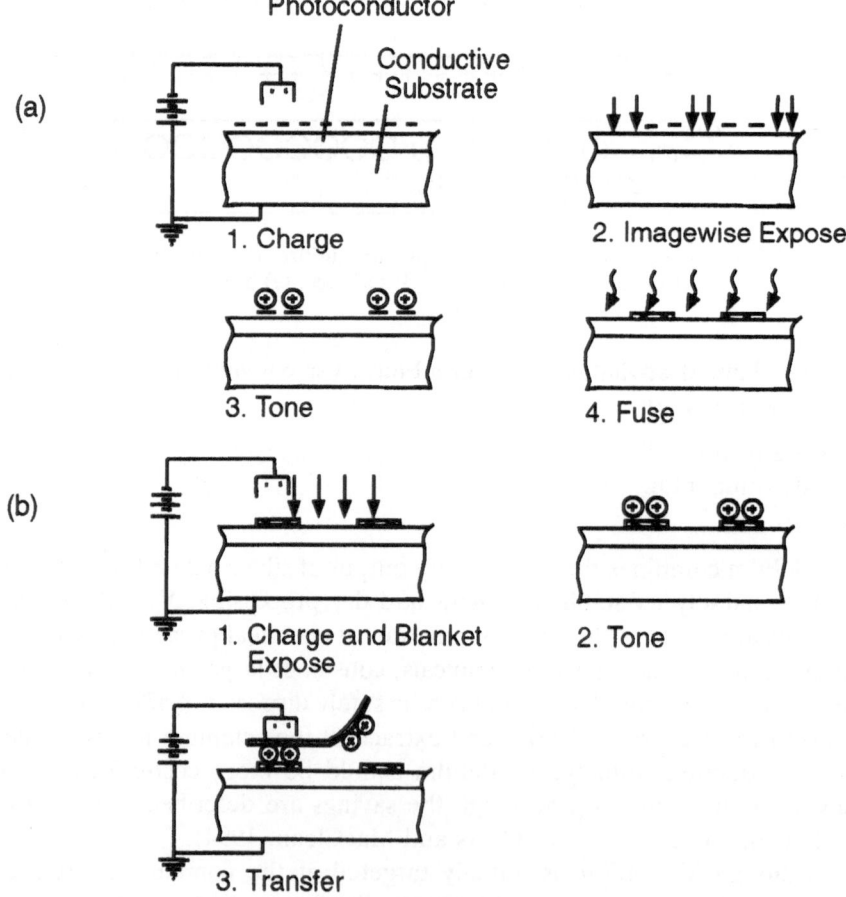

Figure 8.9 Schematic of toner fusion technology: (a) master-forming step, (b) printing cycle.

copy, it is heat-fused directly on to the photoconductor. This becomes the master for subsequent printing runs. For a print run, the master is given a uniform surface charge followed by a blanket exposure to white light (Figure 8.9b). The charge remains in the imaged toner areas because the toner is an insulator. However, it is dissipated on the photoconductor itself (step 1). In step 2, the latent electrostatic image is made visible by toning and the toned image is transferred to paper and fused (step 3). The printing cycle is then repeated. When the printing job is completed, the master is scrolled, rather than cleaned, and reused.

Drawbacks to the basic technology include the need for a separate, blanket exposure for each print cycle. Also, the non-planarity of the fused toner image makes it difficult to clean any residual, untransferred, second-stage toner from the master after each print is generated. As a result, image quality tends to degrade over longer print runs. Nonetheless, this technology has been commercialized as the Xeroprinter 100 by Fuji-Xerox.

8.5.3   TonerJet

TonerJet is a hybrid of laser printing and ink jet printing. Conceptually, it is an ink jet printer in which the ink is a toner. Like ink jet printing, it is essentially a primary printing process in which the toner is fired directly on to the substrate. However, unlike ink jet printing, the toner then requires heat fusing to fix it to the paper.

A simplified TonerJet print zone is shown in Figure 8.10. The dry toner is brought from the toner container into the print zone on the surface of a rotating steel tube, the so-called 'developer sleeve'. During this process, the toner becomes electrically charged. Next, and very close to the developer sleeve, is a control electrode in the form of a thin flexible printhead circuit (FPC) board with many microscopic holes, one hole for each horizontal dot position on the paper. On the other side of the FPC, the paper is fed through. The paper passes on top of a back electrode which is connected to a power source and has a potential of c. 1.5 kV. It creates an electrostatic field between the back-electrode and the sleeve which attracts the charged toner and pulls it through the small holes of the control electrode on to the paper. The key feature of the control electrode is that each hole can be switched on or off, thereby determining whether toner can pass through or not. This is achieved by applying an electrical potential to the ring electrodes around each hole. Thus, when the paper enters the print zone, electrical pulses are sent to the ring electrodes of the control electrode in the correct sequence so that each 'jet' of toner results in a dot on the moving paper to form the image. The paper then continues into a fuser which, as in a laser printer or photocopier, fixes the toner to the paper with heat and pressure (Johnson and Larsson, 1993).

Array Printers AB of Sweden are developing the TonerJet printer.

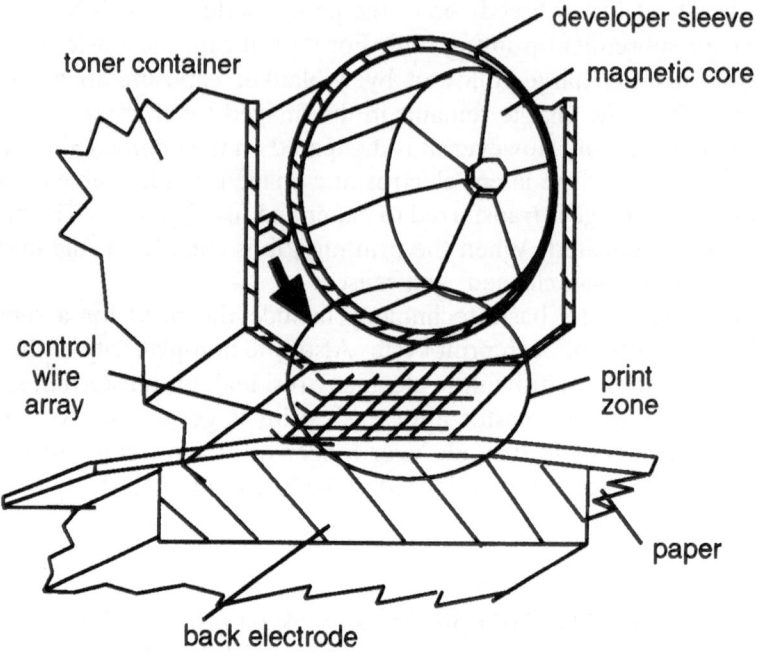

Figure 8.10 Simplified TonerJet print zone.

Experimental printers are claimed to operate at speeds up to 50 cm/s with a resolution of 300 dpi. Initial applications envisaged are label printing, facsimile machines, digital copiers and electronic printing.

8.6   Future prospects

The electrophotographic technologies of photocopying and laser printing are extremely well established, with an installed base of over 50 million units worldwide. Ink jet printing is also an established technology, with over 15 million placements. These technologies are backed by some of the largest companies in the world, such as Xerox and Hewlett Packard in the USA, and Canon in Japan. Such dominance is difficult to break. Therefore, unless any new technology has significant cost and technical advantages over electrophotography and ink jet printing, they are likely to be restricted to a niche market.

References

Anon (1994) New applications drive wide format market. *Printout*, **18**(3), 14–15.
Bugner, D.E. (1991) A review of electrographic printing. *Journal of Imaging Science*, **35**, 377–387.

Caley, W.J. Jr., Buchan, W.R. and Pape, T.W. (1991) Time-resolved charge measurements on ionographic printheads. *Journal of Imaging Technology*, **17**(2), 51–56.

Campbell Scott, J. (1988) *Journal of Output Hardcopy Devices*, R.C. Durbeck and S. Sherr (eds), pp. 263–270. Academic Press, Boston.

Coates, J. and MacClean, J.N. (1994) *Chicago Tribune*, 2 November.

Durchon, P. (1993) Magnetography, a step forward. *Ind. Graphiques*, **417**, 22.

Eltgen, J.-J.P. (1990) Advances of magnetography in very high speed electronic printing. *Proceedings of SPIE, International Society of Optical Engineering*, Bellingham, pp. 104–113.

Ettinger, M. (1994) *Presentation at Villa d'Este, Lake Como, GEC '94*. Also, product literature from Xerox.

Gregory, P. (1991) *High Technology Applications of Organic Colorants*, pp. 207–212. Plenum, New York.

Ishii, S., Higashide, A. Mizoguchi, N. *et al.* (1993) *Nippon Steel Technical Report*, No. 56, pp. 11–14.

Johnson, J. and Larsson, O. (1993) TonerJet – A direct printing process. *IS&T's 9th International Congress on Advances in Non-Impact Printing*, Yokohama, p. 509.

Kubelik, I. (1990) Hard copy and printing technologies. *Proceeding of SPIE, International Society for Optical Engineering*, Bellingham, pp. 45–53.

Lo, A.W.F., Omodani, M., Ohta, M. *et al.* (1993) Image development by liquid toner in ionflow printing. *Journal of Imaging Science and Technology*, **37**(4), 405–410.

Omodani, M., Ohta, M., Fujita, M. *et al* (1993) *NIT Review*, **5**(3), 82–87.

Potter, J. (1987) *Computer Technical Review*, Winter, pp. 102–104.

Skinner, D. (1993) Large format electronic printing. *Oil and Colour Chemist Association Conference*, Slough, UK.

Smith, B. (1991) Printing with electrons. *Byte*, **16**(10), 185–186.

Stobbe, A. (1993) Ion deposition printing. *Flexo*, **18**(12), 76–77.

9   The future

P. GREGORY

9.1   Introduction

Predicting the future correctly is impossible. Nonetheless, it is possible to offer some opinions on future trends in printing and imaging. However, because of the dramatic and rapid changes that have occurred in the last decade or so, such predictions are risky. The degree of risk in predicting the short-term (next five years) future is relatively small, but is extremely high for more longer-term (ten years or more) predictions.

9.2   Short-term predictions

9.2.1   Traditional printing and imaging

Traditional printing technologies, such as offset lithography, flexography and gravure, and imaging technologies, such as silver halide photography, will be around for many years. There are no technologies on the horizon capable of competing with offset lithography or flexography for doing large run printing (millions of prints) at the same cost and with the same speed, quality and reliability. Consequently, newspaper printing will continue by these technologies until well into the 21st century. Similarly, silver halide will remain the dominant photographic technology for at least the short-to-medium term (5–10 years). Although alternative technologies do already exist, they currently have limitations. For example, with electronic photography, it is the poor resolution of the electronic camera, whilst the VerdeFilm technology is restricted to monochrome images. Also, silver halide photography is firmly established with the consumer, i.e. the public, and has billions of dollars invested worldwide, both with cameras and with films and film processing. It will be difficult to dislodge from such a strong position.

One area which is changing and will continue to change is short-run printing. Here, traditional printing technologies such as offset lithography and gravure are being challenged and in some cases replaced by modern non-impact printing technologies such as electrophotographic and especially ink jet. These technologies will make big inroads into short-run printing over the next five years. Indeed, such systems are already on the market. For example, the Stork Truprint system for textile printing (based on ink jet) and the Indigo E-Print 1000 (Barrett, 1994), and the Xeikon DCP-1 (De Schamphelaere, 1994) systems for desk-top publishing (based on laser printing).

9.2.2   Non-impact printing and imaging

The ideal combination of properties for a printer are high speed and high quality combined with low cost. Ink jet printers come closest to satisfying these criteria. Consequently, ink jet printing will become the dominant printing technology for the office and home market. Together with laser printers, they will completely replace impact printers such as dot-matrix (Gregory, 1991). If high-speed page-width array ink jet printers are developed, these will pose a considerable challenge to the photocopying market for precisely the same reasons mentioned above. Until such printers are available, photocopiers will continue to dominate the reproduction of documents, especially black-and-white ones.

As the resolution, speed and quality of ink jet printing continue to advance, it will enter into new markets. For instance, ink jet printers will compete with D2T2 for high-quality colour 'photographic' imaging applications. Large format ink jet printers using high lightfastness colorants (dyes or pigments) will capture a large part of the 'poster' market.

For the storage of vast amounts of data, then optical data-storage techniques will be used. The move to shorter-wavelength lasers will enable even more data to be stored per unit volume.

New synergies will be found between existing technologies (old and new; new and new) and with technologies yet to be invented (Gregory, 1994).

There is no technology on the horizon which will displace the best of the existing traditional impact-printing technologies or the non-impact printing technologies. Indeed, in the longer term, the biggest threat could come from advances in communications technology leading to a new type of culture which could have profound ramifications on the printing and imaging industries.

9.3   Longer-term predictions

9.3.1   Evolutionary developments

Advances in the printing and imaging technologies have thus far been essentially evolutionary. Traditional impact printing and silver halide photographic technologies have progressed by incremental changes over long periods. The advent of the silicon chip and the explosion in electronics

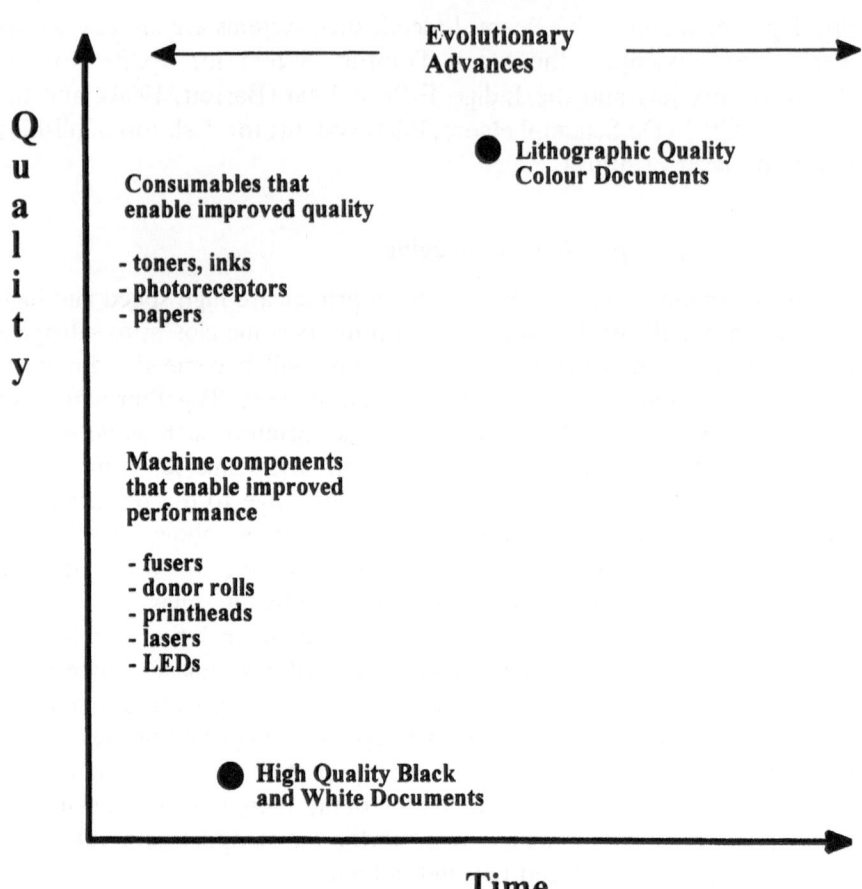

Figure 9.1 Evolutionary advances in non-impact printing.

spawned the current non-impact printing technologies. These are also progressing on an incremental improvement basis, depending upon performance enabling supplies and machine components (Figure 9.1). Recent concerns to conserve energy and natural resources and to protect the environment has led to the so-called 'green' (environmentally friendly) documents. These are produced by recycling waste paper rather than by chopping down forests.

9.3.2 Revolutionary developments

The effect of powerful forces external to the printing and imaging industries could produce revolutionary rather than evolutionary changes in the future. These external forces include the enormous advances in electronic materials, computational power and digital processing, as well as the environmental factors already alluded to. The most important development from these advances will occur in communications. Indeed, the changes will be so dramatic that they will be revolutionary and herald the dawn of a new communications era.

The mainstay of this new communications era will be the information (or electronic) superhighway. This information superhighway will enable anyone with a television or visual display unit to interact with a worldwide source of databases. For example, newspapers and mail would be delivered electronically to your TV screens; books from public libraries would be delivered electronically; and shopping and banking would be done electronically, all from an interactive TV at home (Figure 9.2).

The ramifications to the printing and imaging industries of an information superhighway could be colossal. The first casualties would be the longer-run printing processes, such as offset lithography and flexography, used to produce, for example, newspapers. The demand for traditional hard-copy

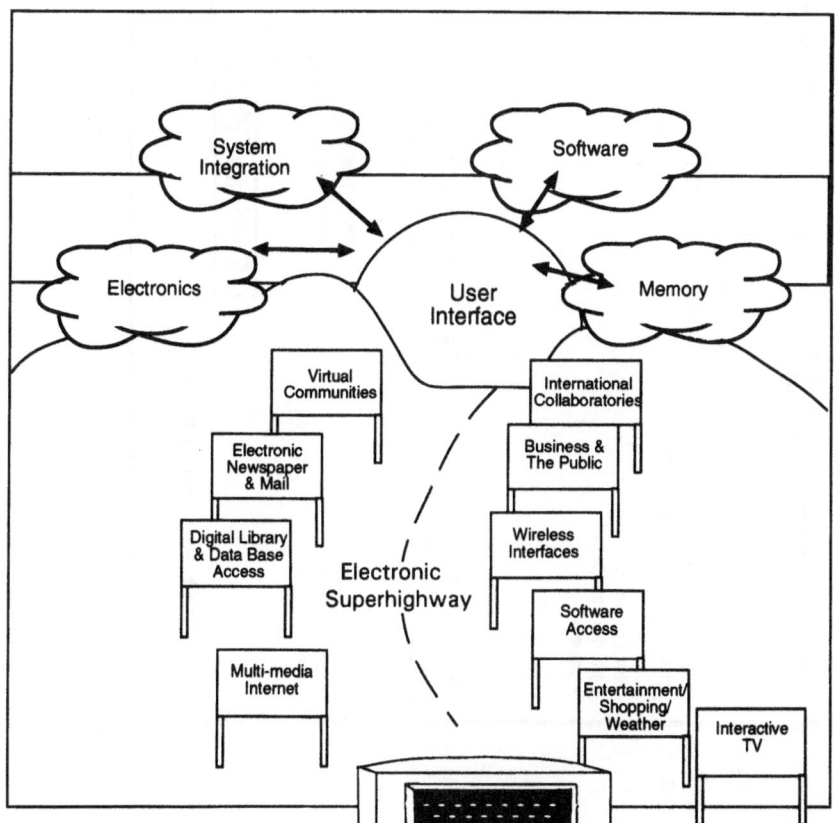

Figure 9.2 The dawning of a new communications era.

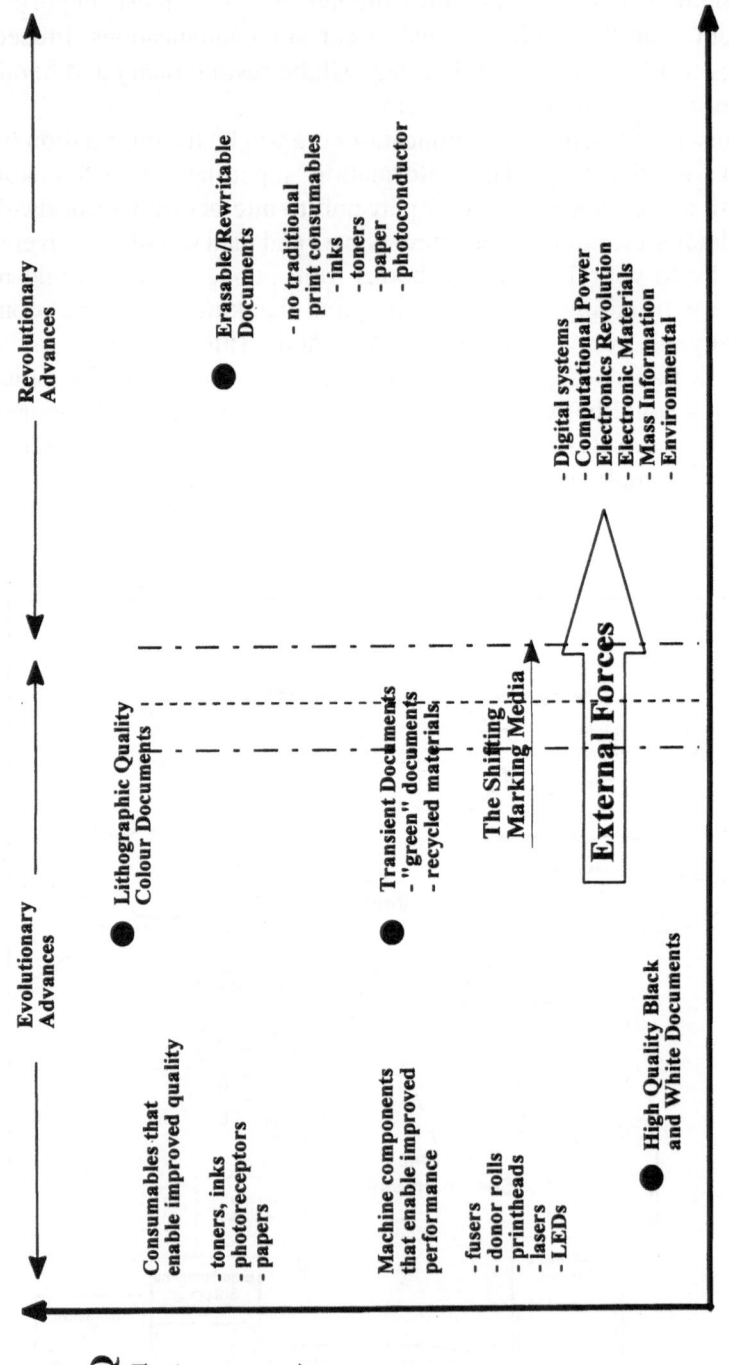

Figure 9.3 Erasable/rewritable documents.

newspapers would decline dramatically. Instead, people would access the newspaper, in full colour, from their interactive TV. Having done so, most people, particularly the older generations, would still prefer to read the newspaper from a hard copy rather than directly from the TV screen. Therefore, the newspaper would likely be printed on to paper using a non-impact printer, probably an ink jet printer. This would prolong the lifetime of non-impact printing, especially ink jet. Longer term, however, the image would be 'printed' on to an erasable/rewritable document. An erasable/rewritable document combines the best attributes of both paper and an electronic display whilst eliminating the drawbacks of both. For example, it would emulate the look and feel of paper, be flexible, lightweight and durable, give high-image quality and contrast, be inexpensive, be colour-capable, have a resident memory, be rewritable and have multidevice connectability. Obviously, erasable/rewritable documents would dramatically reduce the need for non-impact printers (and also paper!) (Figure 9.3).

This scenario may seem futuristic and something that will not happen until well into the 21st century. Not so! Research and development work on the information superhighway is well advanced and embryonic re-imageable documents are already a practical reality. For example, the Expert Pad from Sharp and Notepad 3125 from NCR.

9.4 Conclusion

What will actually happen in the future remains to be seen. However, one thing is certain, whatever happens, the printing and imaging industries are in for an exciting and challenging future.

References

Barrett, J. (1994) 'Woolly inks' transform speed plotting. *Eureka Transfers Technology, April*, pp. 34–35. Also Printing on demand. *Personal Computer World*, March 1994.
De Schamphelaere, L.A. (1994) Digital color presses: applications and technologies. *IS&T's 10th International Congress on Advances in Non-impact Printing Technologies*, 30 October to 4 November, New Orleans, p. 517.
Gregory, P. (1991) *High Technology Applications of Organic Colorants*, pp. 273–283. Plenum, New York.
Gregory, P. (1994) Modern reprographics. *Review of Progress in Coloration*, **24**, 1.

Index